SEEKING TRUTH

Proving Intelligent Design and
Experiencing Our Creator

timm todd

ISBN (Paperback): 978-1-937741-12-9
Printed in the United States of America

Dedication

Dedicated to all the wonderful friends and discussions
that stimulated these topics and sparked this book.

Thank you:

CP, CM, AR, AV, WC, AW, CH, AE, KM, EG, KM

This is what the Lord says:
Stand at the crossroads and look;
Ask for the ancient paths,
Ask where the good way is and walk in it,
And you will find rest for your souls.

~Jeremiah 6:16

Table of Contents

Preface... 1

Introduction 9

Part I - Intelligent Design 13

    A. Vast and Balanced Universe..20
    B. A System that Supports Life 23
    C. The Complexity of Life...29
    D. Math...39
    E. Empty Space 44

Part II - A Universal Creator..53

    A. Life of a Dot..56
    B. Multi-dimensionality of our Creator 57
    C. Moral Code..60
    D. Spiritual Law...70
    E. Losing One's Life to Find It 73
    F. Miracles 77

Part III - The Bible 91

    A. Scientific Accuracy..97
    B. Historical Accuracy...105
    C. Prophetic Accuracy...108

Part IV - Jesus 119

    A. The Universal Creator Made Flesh 123
    B. The Deity of Christ and the Disciples 127
    C. Professor Analogy 132
    D. Sacrifice for Sin 133
    E. Smoking Gun 134
    F. Gravity..137
    G. Faith...140
    H. Free Will..142
    I. Sin and Pain 145
    J. Dying to Sin...147
    K. Power in the Name...150

L. Pizza 151

Part V - Living the Life Intended 155

    A. First Step – Drawing Closer 157
    B. Living Life to Honor Him 160
    C. Refocusing our Priorities 162
    D. Innocence 164
    E. Spiritual Battle 168
    F. Judgment Verses Grace: Eternity 169
    G. Prayer & Transformation 170
    H. Living Abundantly 171

Appendices 177

    A. Transformation of the Disciples 177
    B. Additional Historical Documents 177
    C. Required Elements for Human Life 179

About the Author 183

Preface

Is it absurd to believe in a personified Creator that created the Universe and everything in it, including humanity? Such an idea seems contrary to intellectual thought and science, right? There is no proof. Even if there is some godly being, it is so distant; humanity cannot touch or embrace it in a genuine interactive relationship. If this is you, I, too, once held such views. However, valid evidence supports the opposite position. There is more, so I hope you, the reader, will accept the perspective conveyed herein with an open mind and glean insights to transform your life with peace, purpose, and excitement. You can have it, too. "It is not an institutional model, but a relational model,"[1] in which we may all participate.

Whether we consider ourselves Agnostic, Christian, atheist, Muslim, Jewish, Hindi, or some other religion, how have we sought out truth? Have we studied the *Qur'an*, the *Torah*, the *Bible*, *The Vedas*, the *Tripitaka*, or evaluated many of the world's religions? Have we thought about moral realities or intrinsic dispositions toward certain actions? How do we see events in the world? How do we evaluate history or archeological realities? Are we earnestly listening to our moral compass and programming of our hearts? Have we noticed that the same types of desires that entice us feel hollow once attained? If we are not actively seeking, how do we expect to

---

[1] Hoffman, Buddy, sermon at Grace Midtown Church, Atlanta, Georgia, USA, August 7, 2016.

find?

Life offers dynamic adventure, full of intrigue and raw passion, complimented with patience, peace, and focus that transcends understanding.[2] Rather than continually accepting apathy as religion and fear and uncertainty as life's methodology, let us seize all that life offers. We are all eventually going to die. In terms of the universal ramifications, it does not really matter when; it could be today, or it could be at a ripe old age. Life happens to and around us, whereas our responsibility focuses on our response, through both the things we say and do. Are you ready?

While far from exhaustive, this brief collection of ideas seeks to inspire the reader with an ideological framework to identify his or her beliefs, resulting in daily action. I assert that (1) intelligent design forged the Universe, (2) there is a Designer or Creator (singular or plural), (3) this Creator (who can be called God) is consistent with the deity embraced by Judaism, Christianity, and Islam, (4) the *Bible* is the culmination of the Jewish *Tanakh* and the source for the Muslim *Qur'an*, being much more than just an important historical book, and (5) it is critical we glean a full understanding and view of Jesus to make the most of our lives in this brief period of existence.

This book does not propose any sort of complete iteration or full development of any topic set forth within. It is merely a collection of ideas to facilitate discussion and individual reflection. Any true conversion comes

---

[2] "If anyone would come after Me, he must deny himself and take up his cross daily and follow Me. For whoever wants to save his life will lose it, but whoever loses his life for Me will save it. What good is it for a man to gain the whole world and yet forfeit his very self? If anyone is ashamed of Me and My words, the Son of Man will be ashamed of him when He comes in His glory and in the glory of the Father and of the holy angels," (Luke 9:23-27).

from within. A variety of books delve into several of these topics in much greater detail, supporting a more exhaustive analysis. For readers that may be specialists in an area or have already thought through an idea, these points may seem simplistic and missing, or overlooking, key and important nuances and details. However, for those reading that have yet to make a formal opinion about the ideas expressed herein, hopefully they will stimulate and guide you into a deeper understanding and perspective of objective reality.

As a part of my writing style, I capitalize words when they reflect God or an associated nature of God, whether *Jesus*, *Creator*, *Christ*, etc. While many books use terms like 'intention,' 'love,' 'life-force,' etc., to create a similar connotation, the key distinction is personification of the divine. I use this methodology not only to honor the universal Creator, but also to distinguish between man's subjective or imperfect nature and the supreme nature of this *Entity*.

To try and maintain the flow of ideas, supporting quotes and references drop into the footnotes. Unless otherwise stated, all Biblical and *Tanakh* (consisting of the *Torah*, *Nevi'im*, and *Ketuvim*[3]) references come from a New International Version (NIV) *Bible* and references from the *Qur'an* come from the Qur'an Project.[4]

Increasingly across our world today, feelings dominate and warp a purely critical analysis that rests in facts. In lieu of these qualified truths,

---

[3] The Torah is the first five books of the Bible, The Nevi'im includes both the Former and Latter Prophets, and the Ketuvim (also known as the Hagiographa) includes eleven books of the Old Testament, like Hebrews, I and II Chronicles, Ezra, and Nehemiah.

[4] Qur'an Project, a contextual translation of the Qur'an into the English language. Qu'ran Project, Qu'ran, mobile application, Bedshire, England, 2015.

this subjective reality carries us to a relativism that displaces objective truths, frequently taking us away from a true pursuit of rational analysis. Pure intellectual pursuit requires careful protection against this false relativism. Similarly, our society promotes the notion that science separates physical reality from moral reality, and validates all things in nature and in the Universe. However, nature and the Universe validate science, and science encompasses different aspects of physical reality. There is more to our Universe than a mere three-dimensional physical reality, with moral truths that give weight across dimensions. Additionally, there is another set of laws that shape existence, which are spiritual in nature. Just as true as the former, they require a different means of analysis to explain them.

In our Universe, physicists and laymen alike agree that a complex set of physical laws define our environment, our world, and the Universe. Furthermore, there is reasonable consensus in the scientific method, whereby one creates a reproducible situation to get reproducible actions and observations. My approach follows this same ideology, although my environment is not merely physical, but also moral and spiritual. When an apple falls out of a tree, striking a person sitting below it, it hits him, regardless of his belief in gravity. In fact, it does not matter what this person believes or who he is, as the apple will strike him if he sits between the apple and its previous state of rest on the branch above and its forthcoming state of rest on the ground below. I assert that as with physical laws, there are both moral and spiritual laws that follow this same logic. Whether one believes in them or not, each person is still subject to the reality of each law. Thus, it becomes critical to every person to view life in the context of not only physical, but spiritual and moral understanding, or reality.

A. Physical Laws

Physical law displays its majesty all around us. Palpable to our senses, it transposes itself through sight, sound, touch, taste, and smell. With gravity, an apple falls; with a storm, the wind blows. It invokes emotions, whether grief or happiness. Physical law warms, as well as cools, and may be sharp or soft. We define its truths through science and repeatable actions, as well as observations. While limited in scope and breadth, it frequently purports itself to be absolute reality. With a focus on simply physical laws, we remove moral and spiritual overtones where life may seem devoid of purpose outside of pleasure or subjective pursuits. However, in Chapter 1, we will find that physical reality may be even more of an illusion than the following moral and spiritual realities.

Examples of Physical Laws

While by no means an exhaustive list, a few examples include:

1. An object set in motion will stay in motion, until other forces are set against it[5].
2. For every action, there is an equal and opposite reaction.[6]
3. An object at rest tends to stay at rest, unless there is another force upon it.

---

[5] Commonly referred to as Newton's first law of motion, relating to inertia.
[6] A simplified definition of Newton's third law of motion that as you pull on something, something is pulling on you.

4. Every particle pulls at every other particle with a force consistent with its mass and inversely consistent with the square of its distance, also known as *gravity*.

5. Forces identified across the Universe, including gravity, electromagnetic, nuclear, and strong and weak forces.

6. Energy in a system and the relationship of the energy in that system and systems around it, whether in equilibrium or not, tend to manage toward equilibrium, which also includes uniformity of temperature. The 'loss' of potential energy as *kinetic* is due to things like friction and particle physics of work and heat, as qualified in the four main theories of the Laws of Thermodynamics.

7. When a person is walking in a moving vehicle like a bus, an observer stationary on the ground sees the person's speed relative to the observer. Similarly is the sum of the walker's speed relative to the bus and the bus' speed relative to the observer, as set forth in Classic relativity defined initially by Galileo and refined by Sir Isaac Newton.

8. Special relativity about mass, energy, and momentum causing a bending of the space-time continuum, with gravity being a movement along the least energetic route along a curved space-time.[7]

---

[7] Einstein, as initially published in 1907 and refined through his career, written as (curvature of space-time) = (mass-energy density) * 8 pi G / c4.

For follow-up on certain topics, I list detailed references and encourage those reading to pursue further on points of contention, question, and interest on their own.

Introduction

Racing down the steep descents of Sunshine Canyon, I lean my road bike into each corner and transition my weight from entry to exit to maximize my speed through each apex. Boulder's crisp, blue sky brings heightened awareness, as I head toward the second half of my grueling "seven peaks" training session. Despite the long descent, sweat still drips into my eyes from the prior climb. Cycling is one of the few places where I think clearly and without distraction. It has always been a passion and my peace. As I pedal past sections of sparse trees and green meadows, my heart rate rises, my breathing deepens, and my body sweats, but my mind drifts. Rather than focusing on my heartbeat or pedal stroke, I see the Rocky Mountains and flatlands extending to Kansas with irony, questioning the threads of observable reality around me. I cannot help but feel overwhelmed with gratitude and love, as the beauty unfolds with each intricate turn in the road. I rejoice in the freedom my bike brings.

Cycling is my creative outlet, my tapestry on which I paint not with brush and colors, but pedal-strokes and passion. I feel a symbiosis with peace and purpose that brings clarity. Being able to partake of such wondrous activity fills me with tremendous gratitude that I cannot help but praise and reflect on my existence. My thoughts jump to questions about the reality of our physical existence on planet Earth and the juxtaposition this perspective holds.

I do not believe in *religion* and frequently prefer the authenticity of a sincere atheist to the contrived perfection of many people sitting stagnant in pews of buildings throughout the world. I love the hard questions, the genuine seeking of Truth, and the realization that our own reality is a thin veil of what is really happening in the world. This is true from the politics, economics, and spiritual reality around the world to, frequently, the conversation with our neighbor down the street. The façade of vanity painted over people's true motives oftentimes clouds touted love and selfless pursuit. Even the idea that people, tiny fleeting bags of protoplasm in the Universe, can, through their own actions or works, pursue any form of deity seems trite. We are but dust. The Universe extends around us, yearning for us to see and know *It*. *It* knocks[8], reaching out to us through our lives and in the majesty of the world, even the Universe around us. Our task is to open the door to see and acknowledge *It* with our eyes, hands, heart, mind, and soul.

Life's excitement and color drains without hope. Hope avails us to wake up in the morning and takes us through the day. Like new romance, hope puts a kick in our step and fervor in our stride. However, it matters how we build the foundation upon which our hope rests. Not all sources of hope fill and satisfy equally. My hope rests in my understanding of my place in the Universe, small yet grand. The more we understand our Creator, the more hope, color, and enjoyment, coupled with the less worry, stress, and despair, we experience. This results in greater happiness and

---

[8] "Ask and it will be given to you; seek and you will find; knock and the door will be opened to you. For everyone who asks receives; the one who seeks finds; and to the one who knocks, the door will be opened" (Matthew 7:7-8).

ultimately, peace in each of us. A personified *Force of Creation* exists around and within us, and we all have the capacity to hear and see It, as well as grow in relationship with IT, if we simply slow down and listen. My bike rides afford me the opportunity to do just this and bask in the majesty around me. I wonder how people make it through life, day-to-day, without experiencing it.

I ride back toward Boulder to head up the steep, roughly four-mile Flagstaff mountain, termed "Super Flag," but I could not be further from the pain my pace wreaks on my body; my mind drifts deeper. I wonder why our physical reality seems so shallow and empty, yet purports to be so grand with real, true, universally objective reality hidden in the seemingly sublime. How and why did the Universe come into existence, and why does the astrophysical and molecular composition in which we dwell seem like a juxtaposition of reality? Why do so few people ever discuss or even acknowledge such topics, and why does it seem so *taboo* when they do? Is there meaning or purpose in the Universe that applies to our lives? If there is, it seems critical we uncover our role and position. Looking at the complexity of life and vastness of the Universe, is there more than striving to eat, sleep, work, perhaps workout and, if we are lucky, take an occasional vacation? These simple questions matter to me, as I snap back into my heavy breathing, rocking my bike back and forth out-of-the-saddle, straining up the 11% average grade.

Finally, after another descent, on the way back through town, I ride my bike into a busy intersection and see a car speeding toward me from the cross street; will it stop? Is it fair for me to presume that if my light is green, the driver's light is red? Furthermore, as this driver gets closer to the intersection, I presume that with a red light, his car will slow down and stop, allowing me to continue through the intersection unhindered and without collision. Honestly, I do not even care or think about it. Even though I do not know the driver and cannot see the color of his traffic light, my prior experience confirms a statistical likelihood that when my light displays green, the opposing light shows red. I cannot see this opposing red light, but I trust its color will stop cross-traveling vehicles. My belief plays out into action. This is *faith*, defined as *belief in the unknown, which manifests associated action*. We all have it, so as we delve into the pages of this book, how does your faith play out in your life? What is the foundation of your beliefs? Are all ideas equally valid, or is there objective truth?

Part I

Intelligent Design

Intelligent Design

For millennia, humans knew little about the Universe. Over the last few centuries, we discovered details ranging from nuances of the atom to the profound scale of the cosmos. With each new discovery, we become increasingly aware of the intricate, sophisticated, and fragile complexity of life. Even with the acceleration of our research and understanding, we know only a fraction of objective reality. From the depths of the ocean to the far reaches of space, in every conceivable field of study, new breakthroughs continue to emerge, and this will continue long into the future.

Many people believe the vastness of space and time proves we are spiritually alone and that our lives demonstrate no overall significance. Some time ago, I spoke with one of my friends, Chris, while camping in the mountains of Colorado. The sun faded into the night sky, and we looked up amidst a plethora of stars, overwhelmed by the immensity of space and time. Chris commented, "Amidst these billions of stars and planets, I cannot help but feel spiritually alone. We are so small and trivial in the overall scheme that our lives and even existence cannot show any real significance." He expressed an interesting and widely shared perspective, seeing the Universe as a cold and random place where humanity exists simply by luck, with seemingly no role, rhyme, or reason. For Chris, there is no spiritual or moral dimension beyond the short span of our physical lives.

As we continued to talk over the newly kindled fire, I shared my own perspective that the vast night sky and the intricacies of the cosmos, demonstrate profound design. It is in times like these where my existence on this planet, and in the Universe, seems to hold tremendous value. In this moment, and in similar conversations with other friends, I realized that in most cases, they are unaware that their view is not scientific, but based upon a distinct philosophical position, birthed from a set of false assumptions. Looking at the same scientific facts from a different set of assumptions leads to a completely different world-view. In this chapter, I present scientific and philosophical evidence revealing both design and a propensity for life, pointing us to an *intelligent force* that created the Universe. This, in turn, brings tremendous value to each one of us, and with this, tremendous responsibility.

In the Middle Ages, dating back before Aristotle in the fourth century, the belief that life could *spontaneously generate* from non-life was quite common.[9] Pasteur's experiments (1860s) with a flask of liquid and flies on meat,[10] which many of us remember from school, shut the door on that fanciful belief. Similar to the false notion of spontaneous generation of fruit flies, many scientists expect the Universe to generate life from nothing. Our Universe produced life, where at one time, it did not. The scientific community calls this transition, from non-life to life, *abiogenesis*

---

[9] Aristotle, The History of Animals, Book V, Part I. translated by D'Arcy Thompson, Oxford Clarendon Press, 1910, "So with animals, some spring from parent animals according to their kind, whilst others grow spontaneously and not from kindred stock; and of these instances of spontaneous generation some come from putrefying earth or vegetable matter, as is the case with a number of insects, while others are spontaneously generated in the inside of animals out of the secretions of their several organs."

[10] Schwartz, Maxime. *The life and works of Louis Pasteur.* Journal of Applied Microbiology. 2001, volume 91, issue 4, pages 597–601.

(the beginning and origin of life).

Scientists must acknowledge that when they study *abiogenesis*, they make certain philosophical assumptions that begin well before the introduction of scientific facts. One such assumption is that the Universe is a *closed* or *isolated-system*. These terms refer to the Universe running completely on its own energy, without exchanging matter or energy from outside its surroundings, whereby everything that happens within the Universe originates from its own power. Opposing this foundational assumption views the Universe as an *open-system*. An *open-system* exchanges not only energy, but also matter from outside its surroundings. Both theories lack absolute empirical proof, yet both see scientific evidence based upon the perspective used, making the associated scientific view almost philosophical.

When Newton developed his observations on the motion of bodies with various forces (early eighteenth century), now called Newtonian physics, he viewed the Earth as a closed-system. He observed and developed generalized theories about the behavior of objects within the Earth's atmosphere.[11] Nearly a hundred years later when Einstein came along, he revolutionized the field by viewing the Earth as an open-system within a much larger Universe. Insights derived from his theory of general relativity claim that the mass and energy throughout the Universe cause it to bend and curve. Einstein did not disprove Newtonian physics, but rather contextualized it and provided a much more robust explanation, leveraging additional facts and information. Where my friend Chris, above, and I

---

[11] French, A.P,. <u>Newtonian Mechanics</u>, W. W. Norton & Company, 1971, page 3.

disagree is not that he believes in science and I do not. Rather, Chris accepts an assumption that the Universe is a closed-system, whereas I see it as an open-system. Simply stated, Chris and I hold different baseline views that shape our perceptions of the world. However, science, I argue, still supports the philosophical baseline of an open-system. The discussion that follows highlights current scientific facts suggesting critical factors supporting an open-system, leading us to subsequent views of design and associated realities.

From a *closed-system* approach, the origin of life on Earth must result completely from chance and random circumstances. Inorganic building blocks of life must align in just the right ways, at just the right times, and in just the right locations, to produce life. If the situation is by chance and completely random, this suggests that statistics and probability can be applied to the process as one gauge of the feasibility of this *closed-system* approach. This argument is called the "odds-based" approach and is a good starting point before proceeding with other arguments. Simply stated, the *odds-based* approach highlights the fact that the odds of life emerging randomly in a *closed-system* are so infinitesimal that they move from the realm of scientific fact to the realm of fantasy.

In supporting life-by-chance, leading evolutionary biologist, Richard Dawkins, takes a biased perspective. He states that the Universe began from nothing, adding:

> *"Of course it is counter intuitive, that you can get something from nothing. Of course common sense does not allow you to get something from nothing. That's why it is interesting. It has got*

to be interesting in order to give rise to the universe at all. Something pretty mysterious had to give rise to the origin of the Universe. But that's exactly what's meant by nothing. But whatever it is, it is very, very simple."[12]

In another speech, he expands his argument by numbering the planets in the Universe, which he approximates to one quintillion (one billion, billion). He, then, turns to the odds for life emerging from non-life. At this point, he performs the following thought experiment, "Now, suppose the origin of life, the spontaneous arising of something equivalent to DNA, really was a quite staggeringly improbable event. Suppose it was so improbable as to occur on only one in a billion planets."[13] So for Dawkins who wants to fashion himself the impeccable scientist, it seems as though his starting point and basis for his theory builds upon a highly speculative and unscientific value. He presumes an extremely generous number where spontaneous generation occurs on every 10^9 planets. Unfortunately for Dawkins, more reliable and factually-grounded statistical data exists, which we delve into later. This evidence derives from two distinct perspectives on the requirements for life.

People believing that with enough time, life emerging from nothing is possible tremendously over-simplify the complexity of the situation. Even Dawkins' 10^9 is not enough to foster a razor-thin margin of a vast list of highly sensitive variables, unless we change the definition of nothing to

---

[12] Dawkins, Richard, Public debate between Dawkins and Arch Bishop of Sydney, Australia and Cardinal George Pell for Australian television show *Q&A* on April 9, 2012. Accessed on 10/4/2017 at https://www.youtube.com/watch?v=1J8VvMmV36o.
[13] Richard Dawkins, <u>The God Delusion</u>, Mariner Books, 2006, pages 137-138.

include something. Like earlier views of spontaneous generation, Dawkins espouses falsely suppositious logic. An appropriate analogy places all the raw materials for a car in a huge bag and shakes the bag, until the parts become a Ford Model T or a Tesla X. The odds for something like this occurring are so small, that we must compare it to the quantity of particles in the entire Universe to illustrate the remoteness of the odds, presuming the process either needs an infinite amount of time, which we know the Universe does not have, or miraculously phenomenal luck (which is a bit of an oxymoron). This scenario and its "real-world" counterpart are so unlikely, they should be considered pure fiction and based upon personal bias, rather than fact. If one looks at the facts, the most probable argument explains the Universe has at least a pre-disposition toward order and life, enabled by a methodical intelligence.

A. Vast and Balanced Universe

The sheer scale of our Universe boggles common understanding, and its vast size offers perspective and context, as we qualify the scale for the preceding "odds-based" argument. In failing to properly calibrate the scale, we may miss the enormity of differentiation. Thus, I take a bit of a tangent to help us understand the magnitude of the "odds-based" approach, using the size of the Universe and associated atoms it contains, as our measuring stick.

We understand the geography of our immediate surroundings, like the distance from home to work, measured in minutes, blocks, or miles. We might stretch our scale to states, countries, or continents for school and travel. However, if one were to travel around the world, the circumference of Earth is still a relatively paltry 24,900 miles (39,931 kilometers, or 4×10^7 meters) at the equatorial line. Meanwhile, cosmic distances shatter our typical perceptions of distance and time.

The Universe extends so vastly, we need different units of measure. Scientific circles leverage the Yottameter (Ym, 10^{24} meters), although others frequently use the non-metric light-year, the distance light can travel in a vacuum in one year. There are 106 million light years in a Yottameter. The Great Wall (also known as Coma Wall) is a supercluster of galaxies currently estimated to be 15 Ym (16 million light years) deep, 2.8 Ym (300 million light years) wide, and 4.7 Ym (500 million light years) long. This is 31.59 *trillion* times the distance from the Earth to the sun. If the distance from the Earth to the sun equates the thickness of one piece of paper, it takes 8.6 stacks of paper reaching to the moon to match the size of the Great Wall.[14]

In 2016, *Hubble News* reported that the space telescope identified over one hundred billion galaxies.[15] Many hold billions of stars. Despite the billions and billions of stars and galaxies, researchers estimate there are only 10^{78} to 10^{82} *atoms* in the *entire observable* Universe.[16] Absorb this number. NASA launched *Voyager 2* in 1977, and it has been traveling for over forty years just to get outside our solar system. Our solar system dwells off in the shadows of seeming obscurity, on the fringes of a spiral arm made up of

---

[14] Estimating a ream of 500 sheets of paper to be 5.2 centimeters, with the distance to the moon being approximately 238,900 miles.

[15] Hille, Karl, https://www.nasa.gov/feature/goddard/2016/hubble-reveals-observable-universe-contains-10-times-more-galaxies-than-previously-thought, Oct. 13, 2016, retrieved January 5, 2017.

[16] John Vilanueva, *Atoms in the Universe*, Universe Today, July 30, 2009 [http://www.universetoday.com/36302/].

the billions of stars in a galaxy made relatively non-descript amongst billions of galaxies.[17] Thus, in contemplating the number of atoms in the Universe, this 10^{78} to 10^{82} helps illustrate the magnitude of such a large number and qualifies the outer limits of definition.

Further, many scientists currently accept an age of the Universe of about 13.8 billion years.[18] To help qualify meaning to the magnitude of scale, to support life on planet Earth, every second of every moment throughout this period is approximately 4.35×10^{17} seconds. When we look at every particle in the observable Universe being about 1×10^{80}, and life of the simplest and most basic organism being about 1×10^{140}, with the chances of a planet to create life as we know it being well over 1×10^{1032}, life as we know it requires nearly each particle in the Universe doing something at precise times, in exact locations, throughout all of time. These same scholars miss huge orders of magnitude that show the Universe as extremely effective and systematic. Even through leveraging every second of every year throughout many millennia, seizing every environmental condition at just the right time, the chances for creating human life on Earth still look profoundly efficient. Furthermore, it seems altogether plausible that the propensity and design for life here means life must exist in other areas of the Universe, as well. This illustrates a purposeful

---

[17] DeGrasse, Neil, Welcome to the Universe: An Astrophysical Tour, 2016, stating that according to NASA, scientists estimate that there may be tens of billions of solar systems in our galaxy, perhaps even as many as 100 billion; Livio, Mario, astrophysicist at the Space Telescope Science Institute in Baltimore, April 1, 2014, interview with www.space.com, estimates an acceptable range is between 100 billion and 200 billion galaxies, stating Hubble identifies around 100M but improvements in telescopes will likely double the number.

[18] Cosmic Detectives, European Space Agency, 2013. Retrieved December 4, 2013. http://www.esa.int/Our_Activities/Space_Science/Cosmic_detectives.

causation of events bringing about a physical reality, as well as bringing substance to moral and spiritual realities. The "odds-based" approach requires numbers vast series of magnitudes greater, shedding light on the required logic to support such an argument.

B. A System that Supports Life

Our Universe also balances elements and forces that when combined and sequenced, foster life. Earth's precise location within this vastness may be critical to its ability to foster life. This seeks to qualify *life-as-we-know-it* and is very different from the much more complex examples we will jump into a bit later, separating simple life, like an amoeba, from complex life, like humans. However, as presented in books and television shows like *Star Trek*, there may be non-carbon based forms of life. Life on planet Earth depends upon thousands of extremely delicate variables. These variables demonstrate the anthropic principle,[19] or the cosmological theory that the presence of life restricts the theoretical development of the Universe to processes many scientists feel led to life. I share four examples merely to illustrate the statistical probability of simple life,[20] all of which are but stepping stones for complex human life.

Element Formation - Specific parameters bring about the heavier elements necessary for life. Astrophysicists generally agree that shortly after cosmic inflation and the *Big Bang*, molecular chemistry began with the

---

[19] Anthropic Principle – "the cosmological theory that the presence of life in the universe limits the ways in which the very early universe could have evolved" (Dictionary.com, accessed December 23, 2013). See also: http://www.worldcat.org/title/anthropic-cosmological-principle/oclc/848417743.

[20] A more complete list can be found in the Appendix

formation of hydrogen (75% of matter), then helium (25%), and lithium (<1%). However, after these first three elements, a pause enabled the Universe to cool slightly, slowly forming into stars and galaxies, and the design we currently see in the sky.

Life on Earth requires a very small, yet critical amount of carbon and heavier elements. These form during the compression of the initial three elements due to stellar nucleosynthesis, planetary nebulae, or supernova explosion.[21] A supernova, for example, can be such a powerful galactic occurrence that it outshines an entire galaxy, radiating more energy in a few weeks, than a sun emits over billions of years. As a stellar core collapses, these three: hydrogen, helium, and lithium (and successive) elements compress with such profound force that at the time of detonation, the molecular structures change into heavier elements, which subsequently shoot out into space in a massive, chaotic shock-wave.[22] Each tier of events of elemental formation must occur in very precise sequence, energy level, quantity, and location to bring about the elements necessary for life.

The *Hoyle State* is a theoretical model describing the primordial synergies of hydrogen, nitrogen, and oxygen in stars. Nucleosynthesis of carbon in helium-burning giant stars requires these synergies. Theoretical astrophysicists around the world now accept it as the preeminent theory for how stars produce heavier elements, which is critical for enabling the elements required for life. However, it shows tremendously extreme fine-tuning. If the amount of energy in stars forming carbon and oxygen

---

[21] Grupen, Claus, *Big Bang Nucleosynthesis*, Astroparticle Physics, Berlin: Springer, 213-28, 2005.

[22] Gribbin, JR, Stardust: Supermovae and Life – The Cosmic Connection. Yale Press, 2000.

changes by as little as a billionth of a billionth of the amount necessary to power an ordinary 60-watt light bulb for one second, the Universe could not support life.[23] A higher energy level would prevent the amount of carbon necessary for life; a lower one would cause stars to burn helium into carbon too quickly and prevent the amount of oxygen necessary for life.[24] This extremely fragile event is critical for the formation of heavier elements throughout the Universe and, statistically, shows either phenomenal luck or precise design. This is only one of numerous precarious factors that change the principle defined in the *Hoyle State* and preclude life.[25] If stellar physics were even slightly different, then the elements that make life possible do not exist.

There are hundreds of similar examples, where life exists in the Universe and where the Universe enables life. Without going into each, I

---

[23] One joule (which equals 6.24^{+15} keV) can be defined as the amount of power to light one watt for one second, with a kilo electronvolt being 1.6×10^{-16} joules, and the Hoyle state qualifying the balance between 279-479 keVs (kilo electronvolts), with a mean of 379 keV. Thus, if the Hoyle state differed by 100 keV, also written as 0.000,000,000,000,016,02 joules, the universe would not form heavier elements required for life , Edwin Cartlidge, Carbon's Hoyle State Calculated at Long Last, Physics World, January 3, 2013. See also: Cartlidge, Edwin, *Carbon's Hoyle State Calculated at Long Last*, Physics World, Jan 3 issue, 2013.
http://physicsworld.com/cws/article/news/2013/jan/03/carbons-hoyle-state-calculated-at-long-last
Lee, Dean and Epelbaum,Evgeny, *Foundations of Carbon-based Life Leave Little Room For Error*, Physical Review Letters, American Physical Society, March 13 issue, 2013. This is a peer-reviewed scientific journal considered to be a prestigious journal in the field of physics. Also accessed on 12/14/2017 at https://phys.org/news/2013-03-foundations-carbon-based-life-room-error.html.
24 479,000 electronvolts of three alpha particles on high end and 279,000 electronvolts on the low wend; http://www.dailygalaxy.com/my_weblog/2013/03/the-hoyle-state-physicists-ask-is-the-universe-fine-tuned-for-the-formation-of-carbon-and-oxygen.html, Accessed October 17, 2013.
[25] The light-quark mass looks to be a relative constant with any change in the energy greatly impacting the protons and neutrons, of which they comprise, and the elements and molecules they create.

list a few more. The important idea to glean from these is that the system is profoundly fragile, to magnitudes of order vastly greater than every atom in the Universe.[26] Thinking such odds demonstrate "science" means one lacks an understanding of the issues.

Universal Force – Many leading astrophysicists feel temperatures exceeded $1x10^{30}$ degrees kelvin within the first fraction of a second immediately following the *Big Bang*. Within the first second, these temperatures cooled down to $1x10^{10}$ degrees kelvin.[27] Moments later, particles collided at speeds that we replicate today only in particle accelerator experiments, where scientists see the carnage of collisions producing quarks, antiquarks, and gluons.[28] These experiments demonstrate not only more essential building blocks to matter, but also prove the existence and composition of anti-matter. I will touch upon the novelty of this idea later.[29] However, without the high energy immediately following universal creation, our Universe would not hold the matter, anti-matter, and forces enabling universal formation and life.

Elemental Ratio - Life requires a precise ratio of elemental "ingredients" to make it viable. Life on Earth is largely based on carbon, but also requires hydrogen, nitrogen, oxygen, phosphorus, and sulfur.[30]

---

[26] With $1x 10^{96}$ atoms in the universe and $1 x 10^{1104}$ odds of life evolving, all the atoms may be multiplied itself ten times.

[27] Siegel, Ethan, Beyond the Galaxy: How humanity looked beyond our Milky Way and discovered the entire Universe, World Scientific Publishing Co, 2015.

[28] http://www.particleadventure.org/comp_qg.html, October 13, 2013. The confirmation of quarks and gluons also validated the theoretical Higgs Boson (or "God particle"), a subatomic particle lending support to the modern particle physics theory of the Standard Model, which currently provides a popular framework for types of matter and forces in the universe.

[29] I discuss this on the section discussing $1+ (-1) = 0$.

[30] Due to carbon's ability to bond with so many other elements, it is an ideal atomic

Even minute variance in the ratio of these necessary elements can be critical. For example, while we require .003% of our overall makeup to be iron, just one extra gram will kill us. There are twenty-six elements critical to life that would prevent our existence if their levels in the body deviated by as little as a fraction of a percent. As shared above, on elemental creation, Earth won the lottery and benefits from elemental ratios that enable life.

The Universe contains about 25% helium, but helium is rare on Earth, comprising just five ten-thousandths of a percent (0.00052%) of our atmosphere.[31] This is dramatically outside the norm, which is fortunate. Life on our planet requires oxygen, but helium diffuses oxygen in the body and is therefore toxic in large quantities. If Earth were not an exception to most of the rest of the Universe, conditions on Earth would not support life. There could be, and some argue there must be, similar planets elsewhere in the Universe, but helium levels are just one of thousands of critical criteria.

Earth Proximity - Astrophysicists estimate that in most galaxies, supernovae happen a few times every century. For such an occurrence to seed our planet with heavy elements, scientists estimate it would need to take place within one-hundred light-years of Earth. However, a supernova within thirty light-years would significantly deteriorate our ozone. Thus, for a host planet to obtain the elements sustainable for life, it must occur at

---

structure on which to base life. Scientists do not see many options for an elemental base for life, although a second, much less-likely option might be silicon. NASA has also discovered a very basic bacterium that replaces phosphorus with arsenic. Wolfe-Simon, Felisa and team, *A Bacterium That Can Grow by Using Arsenic Instead of Phosphorus*, Science, June 3, 2011, Vol. 332, Issue 6034, pages 1163-1166;

[31] http://en.wikipedia.org/wiki/Helium, October 13, 2013.

least thirty light years away, but not more than one-hundred. For a galaxy like the *Milky Way*, which is only 100,000-180,000 light years across, these elements must form at a very specific distance and at a critical period in Earth's history.

These are just four examples. We still have yet to consider the possibility of life actually generating from its building blocks. This means that this exceedingly narrow possibility must then be multiplied by the possibility of the chance generation of life. This calculation is the next piece of the puzzle.

One astrophysicist, Hugh Ross, documented over four hundred such variables, and this is still far from an exhaustive list.[32] Each must be balanced precisely for *life-as-we-know-it*. The same astrophysicist estimates the likelihood of complex life being less than one in $10^{10^{32}}$ that all these variables could line up randomly.[33] One might assume that in such an expansive Universe, it is bound to happen in some place at some time; but that assumption demonstrates a misunderstanding of the scale of this number. As such, $10^{10^{32}}$ dwarfs even the total number of atoms in the Universe by significant orders of magnitude. With such a minute chance of actualization, it totally undermines the possibility of life and order arising by chance through randomness. Believing life in the Universe formed by statistical chance is simply misinformed at best, and delusional at worst. The logical explanation qualifies formation of our Universe through an *open-system*, lending strong support for Intelligent Design.

---

[32] Hugh Ross, <u>Why the Universe Is the Way It Is</u>, 2008. See also: http://www.reasons.org/files/compendium/compendium_part2.pdf
[33] Hugh Ross, <u>Why the Universe is the Way It Is</u>, Bake Books, 2008, Appendix C.

C. The Complexity of Life

As recently as the summer of 2016, the scientific community developed a new equation for calculating the probability of abiogenesis on any given planet.[34] They calculate the average (mean) number of events of abiogenesis expected on a planet using (a) the number of building blocks for life available on the planet, (b) the average number of building blocks needed per organism, (c) the availability of those building blocks during a specific time, (d) the amount of time available, and (e) the probability for those blocks to assemble over time. This last one (e) is tricky, with the equation focusing on the probability for the building blocks of life to assemble; it is unknown and the next focus of our attention.

In discussing this element of their equation, Paul Davies, another astrobiologist, explained that were life to be unique to Earth within the observable Universe, the figure should be around one in 10^{24}. Davies generated this number not from an actual analysis of the process involved, but from a simple calculation of one instance on Earth versus no instances on any other planet in the observable Universe.[35] However, our primary interest focuses on whether this figure identifies a value about where we would expect it to be for the generation of life, or whether it identifies a value far lower than would be expected. If this figure falls within the expected range for the probability of the generation of life (or much higher, meaning that life on other planets is expected), this would support (but not

---

[34] Scharf, Caleb and Cronin, Leroy, *Quantifying the origins of life on a planetary scale*, PNAS, National Academy of Science, Vol 113, no 29, 2016. Also accessed on 1/5/17 at http://www.pnas.org/content/113/29/8127.full.pdf.
[35] Davies, Paul, The Goldilocks Enigma: Why is the Universe just Right for Life?, Penguin Press, 2006.

prove) a *closed-system* view of the Universe. Meanwhile, if this figure proves to be much greater than the expected odds-range, it supports (but not proves) an *open-system* view of the Universe.

Fortunately, researchers already studied the probability that the building blocks of life could form into and generate the most basic life forms. Harold Morowitz[36] begins this task by identifying the smallest-known free-living organism, *Mycoplasma hominis*,[37] as the focus of his attention.[38] He isolates only one of the protein bonds contained within this tiny organism. Focusing on this one protein bond, he then calculates the number of possible permutations of amino acid chains that make-up the single protein bond. He calculates the odds for these amino acid chains to randomly line up to form that single protein bond (a very small part of this tiny living organism) as one in $10^{140,}$ at best. At this point, Davies' one in 10^{24} odds are looking pretty low, and highly inaccurate, in comparison.

Building a protein bond is only the beginning and far from the multifaceted and complex requirements for life, itself. We can simplify life down to two basic things: food and reproduction. Living organisms must generate energy to reproduce. Generating energy requires a food source and the ability to convert that food source into energy. With that energy, living organisms reproduce. There are only a few ways that organisms

---

[36] Harold J. Morowitz (1927-), http://en.wikipedia.org/wiki/Harold_J._Morowitz, accessed October 24, 2013.

[37] Hans R. Bode, *Size and structure of the Mycoplasma hominis H39 chromosome Original Research Article,* Journal of Molecular Biology, Volume 23, Issue 2, 1967, pages 191-196,IN3-IN13,197-199

[38] Richard Peacock argues that one should start not with a protein bond that actually exists, but rather with the "simplest theorized self-replicating peptide." But since such a self-replicating peptide is only theorized, Morowitz' approach using the simplest known organism that exists is a better starting point methodologically.

turn food into energy: either like plants (via photosynthesis or chemosynthesis), animals that consume plants, or animals that consume other animals. Of these processes, eating plants or animals is a much less complicated process than photosynthesis and chemosynthesis. As a result, based upon standard evolutionary principles, the earliest forms of life must have been *food-eaters* and not *food-makers*. But, food itself must be a living organism. So, the first food-eating organism had to have access to food, itself a living organism. It is a paradox.

This two-pronged basis of life is the very root of the problem. Even the most basic reproductive system, like RNA (ribonucleic acid), requires a metabolic system (a cell membrane) to function. Similarly, a metabolic system, consisting of proteins and amino acids, needs to reproduce. Finally, these mechanisms must be held together in some way and yet also compartmentalized. This process requires fatty lipids. This paradox lies at the heart of the scientific debate between looking towards RNA in the search to recreate abiogenesis (the RNA-world hypothesis), versus looking at amino acids or proteins as the starting point. Researchers continue to argue that this must move in a step-by-step sequence, while the very nature of the problem is that multiple complex steps must occur and coalesce simultaneously in close proximity. As early as 1971, Monod took this paradox so seriously he argued that because of the low statistical probability of these building blocks forming and coalescing in one "lucky move," its probability was virtually none. Before you object that he performed this research some fifty years ago and science has come a long way since then, Richard Dawkins basically reiterated this same sentiment in 1996:

"But cumulative selection cannot work unless there is some minimal machinery of replication and replicator power, and the only machinery of replication that we know seems too complicated to have come into existence by means of anything less than many generations of cumulative selection!"[39]

So, this illustrates that the odds for the spontaneous origin of life are not simply one in 10^{24}, but significantly closer to zero, and this excludes cosmic factors necessary for life outlined above. Scientists seem to place considerably more faith into their presupposition of the Universe as an *isolated-system*, than the evidence warrants. Current science supports the idea that chance formation is not a possibility for the origin of life, yet here we remain.

Still, science and physical reality provide further clues and evidence pointing towards a possible solution. The physical world provides some evidence that the very fabric of the chemistry of inanimate matter of our Universe actually has a "propensity to form life." In responding to the internal debate between RNA and protein as the catalysts for life mentioned above, an American chemist, Reza Ghadiri, sought to score a point for the protein side of the debate, as the RNA side had been becoming increasingly dominant in the field. Ghadiri and his team took a molecule from yeast, GCN4, which is fairly unique in its composition based on two protein pieces consisting of an identical amino acid chain.[40]

---

[39] Dawkins, Richard, <u>The Blind Watchmaker</u>, W. W. Norton, 1996, p. 141.
[40] Ghadiri, Reza, *A Self-replicating Peptide,* <u>Nature</u>, issue 382, August 1996, p. 525. Available online at: https://www.nature.com/articles/382525a0

They split the molecule into the two mirrored proteins and then cut each chain in half. They then placed these four segments into a solution at room temperature without any reactive agents and mixed.

Although the process was slow, Ghadiri describes that these segments would need to find each other and link up by random motion. The segments did just that and once they had formed a complete template, they began a much more rapid process of breaking down and replicating themselves. Ghadiri describes what happened as follows. "The only way that can happen is if the product was helping its own synthesis." In other words, there is something inherent, even at the most basic chemical level, that seems to propel simple compounds to form more complex compounds, and to replicate and reproduce. Such an internal propensity could not originate in an *isolated-system* solely governed by chance.

Then, in 2015, researchers at the MRC lab in Cambridge made a breakthrough in this area. The popular publication *Science* led with an article entitled, "Researchers may have solved origin-of-life conundrum." This breakthrough found that ultraviolet light combined with hydrogen cyanide and hydrogen sulfide could create the precursors for nucleic acids, amino acids, and lipids. Hydrogen sulfide was abundant on the early Earth and the hydrogen cyanide present might have come from comets that continually pelted the young planet. The authors noted just one slight problem. The catalysts for these different chemical reactions could not have occurred at the same time. They are mutually exclusive. These types of mutually exclusive, external, and environmental factors are common in these abiogenesis studies, especially when layering onto and feeding results from multiple studies.

It was, however, their earlier 2009 study, which is of most interest here. In this study, these Cambridge researchers concluded that there is a "pre-disposition" for the probiotic synthesis of pyrimidine nucleotides. But, the idea of a pre-disposition and undirected matter also seems to be mutually exclusive. How could matter be pre-disposed to certain chemical reactions? This opens up the possibility of an *open-system* concept of the Universe. If some force external to the Universe had pre-disposed matter to form life, then the odds go out the window. The *open-system* receives inputs from outside, in this case, the Universe, establishing a particular order or direction, again in this case, for life. Pre-disposition infuses a targeted, or directed, element to the process, no longer restricted by randomness or chance.

The conditions necessary for life to come about are only the first step. Once the ingredients are present, they still must come together in organic molecules and then proteins. A protein is composed of a complex chain of amino acids that must be in an exact order for the protein to function. The loss of proper sequence may even be harmful. While the simplest protein may contain fifty amino acids, some may require thousands, with a rough average being 288.[41] We return to Harold Morowitz,[42] a leading biophysicist and expert at applying thermodynamics to the origins of life, and his estimate of the odds of one in 10^{140} in forming one of the protein bonds in

---

[41] Charlotte Nyvold, Svend Birkelund, and Gunna Christiansen, The Mycoplamsa hominis P120 membrane protein contains a 216 amino acid hypervariable domain that is recognized by the human humorial immune response, Microbiology, volume 143, pages 675-688, 1997.

[42] Harold J. Morowitz (1927-), http://en.wikipedia.org/wiki/Harold_J._Morowitz, accessed October 24, 2013.

the bacteria *Mycoplasma hominis*.[43] We see such odds in flipping a coin showing heads 466 times in a row.[44] It seems plausible, until one starts flipping coins. While it *may* happen after just 466 flips, even if every person in the world from the beginning of time flipped a coin every second throughout every day of every year throughout their entire lifetime, such probability is still statistically improbable.[45] This calculation comes after the inconceivable odds detailed in the preceding section, making it dramatically less likely from a cosmic perspective.[46]

Many biologists try to dispute the claims of the "odds-based" argument. Attempting to demonstrate how natural processes could bring about life despite long odds, one scientist argues that the basic laws of chemistry and biochemistry lead inevitably to the formation of polymers from monomers.[47] Similar to the results from the studies on probiotic synthesis, an atheistic biologist seeks to refute the odds premise by arguing that abiogenesis is not random, but rather governed by natural selection, "one of the *mechanisms* that drives evolution."[48] Abiogenesis looks at the

---

[43] Hans R. Bode, *Size and structure of the Mycoplasma hominis H39 chromosome*, Journal of Molecular Biology, Volume 23, Issue 2, 1967, pages 191-196,IN3-IN13,197-199

[44] Calculated as 2 for the sides of the coin, multiplied by the square root of 1 x 10140

[45] Assumes 107 billion people since the beginning of time, with an average life expectancy of 50 years (likely high) resulting in 1.5768 billion seconds per life, being statistical odds of 1.687 x 1020.

[46] For additional reading, James Coppedge goes into greater detail on this point in his book, Evolution: Possible or Impossible, Zondervan, © 1973, p 71-79.

[47] Ian Musgrave, *Lies, Damned Lies, Statistics and Probability of Abiogenesis Calculations*, The Talk Origins Archive, December 21, 1998. In it Ian writes, "Firstly, the formation of biological polymers from monomers is a function of the laws of chemistry and biochemistry, and these are decidedly not random."

[48] Richard Peacock, *The Probability of Life*, Evolution FAQ, 2013 [http://www.evolutionfaq.com/]. Abiogenesis was a long process with many small incremental steps, all governed by the non-random forces of Natural Selection and chemistry. The very first stages of abiogenesis were no more than simple self-replicating

creation of life through natural, non-biological chemical processes, forming simple organic molecules into larger, more complex biomolecules to then, eventually, form living organisms.[49] In fact, such counter-arguments indirectly support the idea that there is some sort of innate universal propensity for life. Rather than refuting the "odds-based" argument, they affirm that a *force* brings organization and design to the apparent chaos of the cosmos.

The assembly of DNA (deoxyribonucleic acid) and RNA present an example of this innate propensity. DNA encodes and stores the genetic information required for the development of all living organisms. Images typically depict it as the popular double-helix twisting "ladder". DNA encodes the patterns for RNA, which performs a critical role in protein synthesis, implementing the genetic code stored in DNA. RNA decays quickly outside of cell membranes, yet without cell membranes, RNA cannot develop. Paradoxically, membrane cells cannot develop without RNA to synthesize them.[50] This *chicken-egg* paradox has become such a problem that some scientists, understanding it as evidence for intelligent design and a creating deity, have proactively looked for theories to refute it.[51]

However, recent studies show this paradox may have a simple solution. One experiment shows an amazing propensity for synergistic formation of

---

molecules, which might hardly have been called alive at all.
[49] Kisak, Paul, Abiogenesis, Natural Processes For The Origin of Life, CreateSpace, 2016.
[50] Vincent Moulton, *RNA: Folding Argues Against a Hot-Start Origin of Life*, Journal of Molecular Evolution, volume 51, 2000, pages 416-421.
[51] Shelley D Copley, Eric Smith, and Harold J Morowitz, *The origin of the RNA world: Co-evolution of genes and metabolism*, Bioorganic Chemistry, 35(6):14, 2007. In this article, Morowitz, Smith, and Copley propose an alternative to RNA, as the scientific community struggles with explaining how macromolecular RNAs emerge from molecules available for creation of life on the early Earth.

RNA by seemingly chance processes, utilizing isolated exposure to trillions of random sequences. Among a multitude of random RNA molecules, a few with catalytic function under certain conditions can reproduce enzymatically, amplifying the process from one molecule generation to another. "It is like molecular choreography, where the molecules choreograph their own behavior," writes the organic chemist that developed these findings.[52] The co-author of a similar study writes, "It is amazing that these nucleotides and bases actually assemble on their own, as life today requires complex enzymes to bring together RNA building blocks and to spatially order them prior to polymerization." [53]

These experiments point to a type of natural selection, even an innate programming, at the molecular level. Although some think this phenomenon is evidence against an "odds-based" argument for Intelligent Design, in fact, these findings support notions of reality's inherent proclivity toward life and order.[54] Many atheists look at this as plausible theory for creating life out of seemingly nothing. Unfortunately, such theories show forms of design yet must rely on organization either originating in chaos, or a pre-disposition in the Universe toward design. They confirm that inherent processes, not random chance, make the conditions for life possible, and they demonstrate our Universe is ordered and structured, not random. Again, it points not to a *closed-system*, but an

---

[52] Southerland, John, from the University of Manchester, as quoted in Matthew W. Powner, Beatrice Gerland & John D. Sutherland, *Synthesis of activated pyrimidine ribonucleotides in prebiotically plausible conditions*, Nature, Vol. 460, May 13, 2009.
[53] John Toon, *New Study Brings Scientists Closer to the Origin of RNA*, Journal of the American Chemical Society, December 14, 2013. Quote from Brian Cafferty, co-author of this study.
[54] See Deamer, Dave, Calculating the Odds That Life Could Begin By Chance, Science 2.0, April 30, 2009.

open one.

This evidence supporting the pre-disposition for probiotic synthesis and RNA creation and choreography holds profound impact and significance. Random chance development eventually evolving into early life forms requires acceptance of the *isolated-system* view of the Universe. However, this propensity of *nothing* towards *something*, building into more complex molecules and self-replication, is simply not possible with the isolated, or *closed-system*, framework. Similar to Richard Dawkins, Stephen Hawking, another adamant atheist and world-renown theoretical physicist and author, believes, "The Universe can spontaneously create itself out of nothing."[55] However, as we have seen, this cannot occur in a purely closed or *isolated-system*. As science shows, and we just highlighted, an *open-system* must occur for any pre-disposition toward Creation. An *open-system*, with options for external energy and matter, fits squarely within these recent scientific discoveries. Since science identifies evidence that physical reality receives direction and 'programming' from an *open-system*, meaning outside the Universe, then it also offers a reasonable framework to support spiritual reality, with rules that impact such reality within consequence in our daily lives.

At this point, nothing has been said about the nature of the *open-system*, whose energy might impact our Universe and hold responsibility for the tendency towards order and life that is inexplicable in an *isolated-system* Universe. However, such order and intentionality points toward a designing

---

[55] Hawking, Stephen, *Questioning the Universe*, TED2008, TED conference, Monterey California, February, 2008. Accessed at
https://www.ted.com/talks/stephen_hawking_asks_big_questions_about_the_universe

force with foresight and intention. If there is an inherency of life and a proclivity toward order, it begs the question of what *force* or *forces* brought about such a design, and is there a Designer or are there Designers?

D. Math

Mathematics offers one of the clearest, as well as oldest, examples of universal order, with its calculations explaining truths and even artistic organization. Math presents not only a means of quantifying many scientific observations, but also articulates the order and design of our world. We see more than random chance in the language of equations. We see an organization and consistency that overcomes the idea that we exist simply through a long line of random causes and effects, and points us in the direction of an ordered Universe and, perhaps, a Designer or Designers.

In 2001, a team of mathematicians sought to expand proofs on the *Fundamental Theorem of Algebra*, dating back to the 1800s. In building upon the idea that every complex polynomial of degree n has n roots in the complex numbers, they considered harmonic polynomials and unwittingly solved a theory that astrophysicists use to describe gravitational lensing.[56] Gravitational lensing is a phenomenon in which a star, galaxy, or other source of light is deflected by the mass of a huge object or set of objects,

---

[56] Khavinson, Dmitry, From the Fundamental Theorem of Algebra to Astrophysics: A "Harmonious" Path, May 12, 2014. Sourced:
https://www.researchgate.net/publication/255635963_FROM_THE_FUNDAMENTAL
_THEOREM_OF_ALGEBRA_TO_ASTROPHYSICS_A_HARMONIOUS_PATH.
Download a copy, as of December 2014:
http://shell.cas.usf.edu/~dkhavins/files/khavneumann_082807fin.pdf

resulting in the observer seeing multiple images of the same light. While Newtonian mathematical theories predicted such a *bending of light* and Einstein's theory of general relativity allowed the concept to develop further, it was not until confirmation by an astrophysicist from the University of California at San Diego,[57] that the math brought clarity to the long-observed phenomenon. The fact that mathematicians predict a phenomenon as complex as gravitational lensing demonstrates the inherent order of the Universe.

The *Pythagorean Theorem* offers another great example of how seemingly simple observations in everyday life fit into a qualified set of laws, or rules, that can be qualified into a consistent calculation.[58] According to the theorem, with credit dating back to a Greek mathematician, Pythagoras, in the sixth century before Christ, any triangle with a right angle (90 degree corner), the side opposite the right angle equals the square root of the sum of the square of both adjacent sides ($c^2 = a^2 + b^2$). This applies to triangles of any size with any set of angles in the remaining two corners. All right-angle triangles fall into this rule, without exception. This theorem shows order and organization, not just in consistency of similar geometric shapes, but in the mathematical expression of them. It shows *design*.

Another profound example of the mathematical order in the Universe is the *golden ratio*. Two numbers (*a* and *b*) are said to be in the golden ratio when their sum (*a + b*) has the same proportion to the greater of the

---

[57] American Mathematical Society, *Where Mathematics And Astrophysics Meet*, ScienceDaily. ScienceDaily, June 6, 2008.

[58] For a detailed analysis of the many nuances of the Pythagorean Theorem, read: Posamentier, Alfred, The Pythagorean Theorem: The Story of Its Power and Beauty, 2010, Prometheus Books, pages 211-234.

numbers (*a*) as the greater has to the lesser (*b*).[59]

$$\frac{a + b}{a} = \frac{a}{b}$$

This occurs when *a* is greater than *b* and *b* is greater than zero, and is also denoted with the Greek phi (φ or Φ). The following diagram characterizes it geometrically:

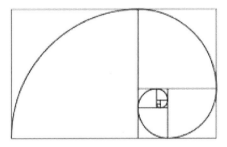

The *golden ratio* interests scholars, as it occurs commonly throughout nature. We see this design in the formation of many seashells, the spiral arms of galaxies, crystal geometry, leaf formation, and in the perfect aesthetic square used throughout the history of design, including in the *Parthenon* in Greece. Nature itself seems to follow an inherent mathematical pattern.

One of the more well-known theories that demonstrates an underlying beauty and order to our Universe is Einstein's theory of *general relativity*, which describes relationship between mass and energy. While this also points to an *open-system*, it helps us ascertain many of the rules, or laws, that

[59] Livio, Mario, The Golden Ratio: The Story of Phi, The World's Most Astonishing Number. New York: Broadway Books. 2002.

govern the Universe's earliest formation, which helps us see the artistic expression, complexity, and simplicity that brought us all into being. Einstein's theory can be summed up in the simple $E=mc^2$, where E is the energy of a physical system, m is mass, and c is the speed of light in a vacuum. In it, we find an explanation of such phenomena as the space-time continuum, gravitational lensing, and the gravitational field of a mass in motion, with gravity being a form of distortion of space-time. "The point is, it's really very simple," says Bill Murray, a particle physicist at the CERN laboratory in Geneva. "But what it embodies is a whole new way of looking at the world, a whole attitude to reality and our relationship to it. Suddenly, the rigid unchanging cosmos is swept away and replaced with a personal world, related to what you observe. You move from being outside the Universe, looking down, to one of the components inside it."[60] Again, it carries tremendous support for an *open-system.*

Further evidence proves the laws intrinsic within Einstein's math with observational science. From the Hubble Space Telescope (and mirrored in the results from the Chandra X-ray Observatory and the Sloan Digital Sky Survey), researchers focused on a perceived 'blank' spot in space and not only identified an additional super massive black hole (weighing in at more than 1 billion of our sun's), but they identified the rippling of gravitational waves, resulting from the possible release of massive energy (theorized to result from the collision of two galaxies and merging of two black holes

---

[60] Moskowitz, Clara, *The 11 Most Beautiful Mathematical Equations*, <u>Live Science</u>, January 29, 2013. http://www.livescience.com/26680-greatest-mathematical-equations.html. Retrieved March 2015.

into this one super black hole).[61] With this release of energy, researchers identified clear warping of light around the fabric of time and space, or gravitational lensing. This offers additional observational support for math's equational science, specifically $E=mc^2$, again qualifying the organization and design found in the Universe, vastly contrasting false notions of pure randomness.

The *theory of relativity* also indirectly supports notions of time in universal Creation, as it only ascribes meaning to objects traveling at, or slower, than the speed of light; thereby, giving rise to new theories around time for energy in a system with motion greater than the speed of light. As such, it is interesting what scholars will say in the future about *The Big Bang* and time, as our assumptions have been based upon a limited view of the universal event being slower than the speed of light, when we now know the initial events must have been much faster than this value. This means that the early expansion of the Universe occurred considerably more quickly than the dates we commonly see published. As shared earlier with theories on the immense heat and energy in the earliest stages of our Universe and its expansion, what looks like billions of years now, with this notion, could have been vast magnitudes quicker, whether millennia, years, days, or even seconds. Einstein's findings show a theme for behavior across the observable Universe, a series of relationships that define an orderly expectation around behavior of the heavenly bodies.

---

[61] Kruhler, T, et al, *The supermassive black hole coincident with the luminous transient ASASSN-15lh*, Astronomy and Physics, volume 610, February 13, 2018, delving into details published on www.hubblesite.org, accessed on March 24, 2017 at http://hubblesite.org/news_release/news/2017-12 and https://www.newscientist.com/article/2125769-stray-supermassive-black-hole-flung-away-by-gravitational-waves/

Events in the Universe, like those predicted by the *theory of relativity*, happen according to a rationale that can be calculated and observed. They demonstrate a sophisticated set of laws from the sub-atomic to universal scale, including simple geometry we see and take for granted every day. The fact that we observe consistencies across these areas is interesting, and the language of mathematics that predicts and articulates them presents a strong argument for the inherent order of the Universe. Think of theoretical mathematics and observational astronomy as participants in a study where two groups work in isolation with very different tools and reach the same conclusions. The ability to arrive at the same conclusion characterizes an underlying organization and structure, resulting in an undeniable fingerprint of design in the Universe.

Science makes observations and predictions about the Universe. Some suggest the fact that the world is knowable at all is evidence of some greater *Mind* at work.[62] The fact is, there is fundamental agreement between atheists and proponents of Intelligent Design on this issue: we live in an ordered and observable Universe that seems to be inherently knowable and programmed for life that can know it. The question we are left with is: *who* or *what* programmed it?

E. Empty Space

Many people assume the Universe is a random and chaotic place, but in fact, it is fundamentally ordered. Have you ever wondered whether reality is

---

[62] Alvin Plantinga, Where the Conflict Really Lies: Science, Religion, and Naturalism (Oxford: Oxford University Press, 2011).

something entirely different from what we think? In the movie, *The Matrix* (1999), the main character may take the *red* pill and see the objective reality and battle the matrix enemy, or take the *blue* pill and simply go back to his own *subjective reality* of living inside the dream created by the matrix. What similar assumptions do we each make about our physical reality? Our subjective view of the Universe ultimately limits our immediate sensory experience. Many people presume the matter of the Universe is physical, containing both material and mass. Hopefully, the chair where you currently sit bears your weight, being solid, and your house rests on a firm, solid foundation.

As we explored previously, space is filled with over one hundred billion galaxies. A single galaxy may hold a hundred billion stars, many composed of a solar system with planets. My point is there is a lot of *stuff* in the Universe. However, let me delve deeper to bring greater context. As we know from our neighboring spiral galaxy, *Andromeda*, which is relatively close at 2.4 million light years in distance,[63] the distance between the *Milky Way* and *Andromeda* is taken up by stars and other celestial bodies, but mostly by empty space, which may be considered *dark matter*. Well over 99.99% of this spatial distance is empty space, being a vacuum of extremely low particle density. These two galaxies even both reside in the same cluster. Galaxies from one cluster to another can be significantly further apart, with super voids up to 4.6 Ym (489 million light years) almost common. Compared to the mass and distance associated with the surface

---

[63] Ribas, I., First Determination of the Distance and Fundamental Properties of an Eclipsing Binary in the Andromeda Galaxy, <u>Astrophysical Journal Letters</u>, 2005. http://iopscience.iop.org/journal/2041-8205.

area of a galaxy, the distances between them dwarfs each.[64]

As another means of qualifying these distances, proportionally, if our sun shrank to the size of a golf ball in the center of New York City, the nearest star, *Proxima Centauri*, which is 4.2 light years away, resembles another golf ball in Chicago, 789 miles away (1270 kilometers).[65] This leaves tremendously vast amounts of the Universe without dense matter from planets, suns, moons, etc. On the cosmic scale, such distances may be dotted with a few celestial bodies, but even in a solar system, over 99.99% of space is empty.

Even when we delve into the physical reality of a planet, like ours, the less than 0.01% of what we think of as celestial matter, is also not as *solid* as it seems. The Earth has trillions and trillions of molecules, each composed of two or more atoms held together by chemical bonds. The atom is, in some ways, like its own miniature solar system, with electrons flying around a nucleus composed of protons and neutrons.

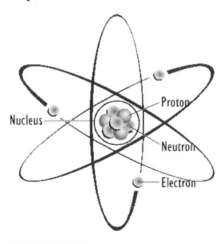

---

[64] Galaxies range from 1500 to over 300,000 light years in diameter and can contain over 100 billion stars (Seeds, Michael, <u>Horizons, Exploring the Universe</u>, Wadsworth Cole, Pacific Grove, CA, © 2002, p.6)
[65] "Except for the widely scattered stars and a few atoms of gas drifting between the stars, the universe is nearly empty" (Seeds, p.5)

If we multiplied the size of a hydrogen atom by one trillion, its nucleus would be about the size of a grape seed, or .16 cm in diameter. We would find its electrons approximately 400 meters away. An atom is also well over 99.99% empty space. Due to the required dissipation of heat, even sub-atomic particles, such as quarks and their gauge particles, gluons, can be calculated to be at least 90% empty space. Even the smallest building blocks of the Universe seem to be made up of something smaller. As one delves even further, some scientists ascribe to a theory that strings comprise a neutrino. A string is thought to be associated with the theory of quantum gravity, whereby a string is a Planck length, which is 10^{-33} centimeters. This equates to a millionth of a billionth of a billionth of a billionth of a centimeter. Planets, moons, and other forms of celestial matter compose mostly *empty* space, too. It leads us to question the very composition of the Universe, but this logic goes even further, making physical reality seem more like an illusion. The thing enabling matter to meet our perception of physical reality is electromagnetic force; without it, our hand could pass through the glass we hold.

Yet, even electromagnetic force supports the same driving narrative of an open or non-isolated system in the Universe. In a closed or *isolated-system*, we again think of Newton. Newton's first law of motion states that an object in motion tends to stay in motion, unless an external force is applied to it. This seems to be true in many situations, except when looking at atomic particles. An electron does not slow down, despite numerous forces upon it, from gravity to other electromagnetic forces that keep them circumventing the nucleus. While Newton's laws of motion apply to other objects, if they applied to particle physics, our Universe would explode in a chaotic display of particle annihilation. While temperature makes atoms slow (or speed up with high temperatures[66]), theories of quantum mechanics explain why gravity from the nucleus does not eventually absorb electrons, as well as why electrons do not shoot out, in tangent, to the plane of motion, away from the nucleus. Because the electron is said to already be in its lowest energy state, it cannot lose energy. The discussion quickly becomes philosophical and is even called the *Heisenberg Uncertainty Principle*,[67] because what *should* happen does not occur, scientists do the *unscientific* and come up with explanations to justify it. This example of quantum mechanics further supports an open-system, superseding the findings by people like Isaac Newton. Again, Newton is not wrong, but evidence further illustrates how the Universe requires external interactions of information, energy, and material transfer.

Furthermore, even the scant particles comprising our *solid* physical world live a precarious existence. In quantum field theory, scientists postulate that there is a near-equal amount of matter and antimatter in the Universe.[68] When a particle encounters an anti-particle, both disappear. For example, shooting a hydrogen atom (+1) into an anti-hydrogen (−1) results in a release of energy and the elimination both atoms. One might argue that since $1 + (-1) = 0$, the Universe ought to be cannibalizing itself back into

---

[66] Due to the temperatures in the early universe, particle speeds greatly differ from those seen today, meaning things happened more quickly, meaning what we perceive as history could have happened dramatically faster than commonly accepted. I do not delve into "modern" calculations of dating, with the many flaws around isotopic decay rates, but again it follows a misleading narrative, with ages being much more recent than

[67] Kumar, Manjit, Quantum: Einstein, Bohr, and the Great Debate About the Nature of Reality, Icon Books, 2009, pages 282-289.

[68] *Smidgen of Antimatter Surrounds Earth*, Seeker, 2011. http://www.seeker.com/smidgen-of-antimatter-surrounds-earth-1765364934.html#news.discovery.com.

nothing as quickly as it is formed. Nevertheless, here we remain. Again, theories of quantum mechanics coupled with mathematical proofs offer rationale that show the Universe operates by rules to allow itself to behave in what appears to be, on the surface, very counter-intuitive ways, lending further support to the *open-system*.

All this begs the question of what really causes matter, mass, and our existence in a Universe built with, well, *nothing*. If our Universe is mostly empty space and the matter it contains risks full particle annihilation back to emptiness, what keeps all these particles and associated forces in such amazing, eloquent, balance, continuing in seemingly tireless perpetuity? In addition to gravitational, nuclear (or strong and weak forces), and electrical and magnetic charges, we see seemingly complex sets of laws maintaining our existence. We realize that without these rules, as quickly as the Universe formed, it could be cannibalizing itself with antimatter back into *nothing*. The emptiness and precariousness of physical reality is a kind of parable that should point us beyond what we see. We move from a purely physical view of reality to a moral and spiritual reality perspective. Our sense of what is real and solid comprised in our physical reality is an illusion of underlying moral and spiritual realities. If we take the entire Universe, place it in a bag and shake it up, everything in it would simply disappear. Something, nevertheless, has ordered it and is holding it together. We perceive only a little sliver of it.

None of us see objectively. A different perspective adds or removes complexity, changing the way we see and understand. Our subjective perception limits our ability to know perfectly. Our daily lives typically focus on the bare essentials of survival, yet we frequently fail to see how our existence fits into an overarching purpose. We see the world as a static, quiet, two-dimensional painting, but it is a multi-dimensional expanse of action, color, sound, and energy. This is an important point that should motivate us to look at our Universe from an active, rather than passive perspective, as it bodes the question, "Why are we here?" It escalates our thinking to a philosophical, scientific, and spiritual plane and leads us to the possibility that some *Being, Force,* or *Propensity for Creation* made us, or allowed us to be made.

*   *   *

In this chapter, I argue for an *open-system* view of the Universe, rather than a closed or *isolated-system* view. This means some form of reality exists outside and beyond our Universe. It implies moral and spiritual realities that emerge not from religious texts or ideas, but from observations within the scientific communities. Like textbook theories in other disciplines like finance and economics, these scientific communities must perform their work from a *closed-system* perspective, relying on math and logic that relies upon and proves an open-view system. Current science offers a limited purview of any reality outside our Universe. Thus, it begs philosophers, not physicists or cosmologists, to ask about the nature of objective reality and metaphysics. With evidence from many of the latest scientific advancements, our Universe shows inherent design and an inherent tendency, or pre-disposition, toward life. As one delves into the *'odds-based'* approach to creation of the Universe, notions of random chance become not only suspect but unscientific, with greater support for the propensity for order and design, curtailed through an *open-system* and by *Intelligent Design*. We see math and quantum truths integrated into the fabric of the Universe. This leads us to identify an *intelligence* or *being* that generates this inherent propensity toward life. This suggests a Designer.

Scientific analysis provides an ever-growing, expansive set of reproducible data limited to physical reality, which we use to draw subjective conclusions. It is nothing more than a means, or lens, with which to view the world. It only describes certain types of information within this physical reality. W may allow one view of the world to bias our reality, or we may use it as a tool, with time and purpose. It is critical that we expand our analysis to moral and spiritual realities, in order to better discern objective truth. In the following chapters, I demonstrate that once we look, we begin to understand and even interact with realities other than the physical and with a Designer who ascribes meaning and purpose to each of us.

Part II

A Universal Creator

A Universal Creator

From Part One, we understand the Universe holds an intelligent propensity for life with design and a Designer. Part Two seeks to bring credence to this creating *Force*. As a *Force* of universal Creation with a predisposition toward *Intelligent Design* and the creation of life, another word can also simply be 'God'.[69] At this point, this *Entity* may be singular or plural and may be personal or aloof. Several ideas support spiritual and moral realities that help us to define *Intelligent Design* as a form of universal Creator and Deity.

People holding to the scientific method as the perfecter of truth typically seek a proof of flawless reproducibility as a means of qualifying science. Additionally, this qualification of science frequently equates their interpretation of all aspects of truth. While this can be the case in a multitude of situations relating to physical reality, there is also much science and truth that exists outside a clean perspective of the Universe that falls neatly into this box. Moral and spiritual realities typically do not fall neatly into such category, as to be easily measured by the scientific method. However, for the sake of argument, this chapter will 'prove' aspects of moral and spiritual reality through reproducible actions and events. While

---

[69] Merriam-Webster Dictionary, *God*, defined as, "the supreme or ultimate reality: such as: the Being perfect in power, wisdom, and goodness who is worshipped as creator and ruler of the universe," accessed on 3/14/2018 at https://www.merriam-webster.com/dictionary/god.

true scientists will likely mock their perception of lack of process, or diligence in the scientific method, they may be locked into a limited perception of their own understanding. Not all truth fits into such a simplistic box and limited lens of validation.

A. Life of a Dot

While it is good to try and understand the universal Designer, one must first come to terms with the impossible totality of this pursuit. Our perspective is inherently limited. As stated in one religious text, "The Almighty is beyond our reach and exalted in power."[70] As the created, we cannot grasp the depth or breadth of a Creator.

Consider a dot of ink on a piece of paper. A dot is just a one-dimensional object, lacking depth, breadth, and width at a static energy. From the one-dimensional view of the dot, even the majesty of the Universe resembles a dot. When looking at a line, all a dot can *see* is another dot. Its limited dimensionality cannot grasp the length of the line. It is constrained within its one-dimensional view. Now, take the line. For the line, everything else resembles either a dot or a line. When the line views a square, it cannot notice the breadth; to the line, the square just looks like another line. Meanwhile, for the square, when it views a cube, all it can see is one side; there is no depth, and the cube just resembles another square. The square can see a dot; it can see a line, and it can see another square, but it cannot see a cube. The same is true with a cube; it is limited to its three

---

[70] "The Almighty is beyond our reach and exalted in power; in his justice and great righteousness, He does not oppress" (Job 37:23).

dimensions and views all things in these three dimensions, down to a dot. So it is with humans.

When we look at the *Force* behind Creation, a multi-dimensional *Being* (or collection of *Beings*), all we see is a limited view of our image. While humans may be multi-dimensional entities, our physical senses of sight, sound, hearing, taste, and touch are typically limited to just three-dimensions. When we look at our Creator, all we see is our limited image and perception. Although we live in a multi-dimensional Universe, we perceive reality from the bias of our interpretation, our senses. We tend to view our Designer through the lens of limited dimensionality and perspective, which places *It* into a box that limits *Its* true nature. While discoveries and theories of science, such as particle physics, push this boundary, a limited dimensional view impacts not only how we see this *Entity*, but how we interact throughout our daily lives. Our Creator exists and dwells in a reality beyond what we could possibly understand, despite our many vain attempts to do so.

B. Multi-dimensionality of our Creator

A framework of three dimensions typically defines sensory information, which provides input into the world around human understanding and experience. This also limits our understanding. When most of us think of time,[71] it is typically articulated as a linear measure, or line. Past events are further along one end, while future events sit on the opposite end. The

---

[71] People in non-equatorial regions and third world countries frequently view time not as a linear measure but as more of a socially and environmentally determined abstract or relative state.

present falls somewhere between the two:

<----- PAST --------------- PRESENT --------------- FUTURE ----->

Similarly, we think of distance as another line, typically following a geographic path across the surface of the Earth. If we were to graph both together, it resembles a matrix, with time being across the 'x' axis, and distance being along the 'y' axis. To travel a distance or move in time, however, a third variable is required: *energy*. When we add in energy, this square becomes a cube, with the 'z' axis, or three dimensions.

Humans represent a single dot, or point, in this cube, comprising a point in time, geographic location, and the energy we exert into the system. However, our Creator consumes every point in the cube simultaneously. *It* is everywhere at the same time, in the *past*, as *It* is in the *future*, and in the *present*. Another way to look at this concept is that the Creator of the Universe offers three dimensions to our three dimensions,[72] making this *Force* a multi-dimensional being. Like the dot looking at the line, let alone the cube, despite our pride and vanity otherwise, if we come to really think about it, we must simply be in awe of the nature of our Designer. Somewhat ironically, many astrophysicists now feel ample evidence to support claim of ten dimensions[73] to help explain our existence and Universe.

---

[72] I am not sure if this makes God a six dimensional being (3+3), a nine dimensional being (3*3), or a twenty-seven dimensional being (3³). ...from a spiritual perspective, it is irrelevant.

73 Rob Bryanton, Imaging the Tenth Dimension, Trafford Publishing, Bloomington, IN, 2007.

The fact is, we often run into the limits of our understanding, even within our own familiar dimensions. I am an avid cyclist and try to get out regularly for a ride. As someone who has ridden over 250,000 miles, around major cities in several countries around the world, I have seen and felt harassment on a variety of levels. Friends have been hit and killed by motorists, and I have been hit four times. People in cars have hunted me with a paint gun, spit on me, swerved toward me, and thrown a variety of objects at me, almost killing me several times. While I am pretty immune to many driver misdeeds, I share one story for perspective on the subjective nature of our limited understanding.

One gorgeous sunny morning, I jump outside for a long ride. I wear black shorts and a black jersey, with a blue helmet. As I live on what is normally a busy two-lane (each way) street, I must brave traffic for a few blocks, until I turn into a neighborhood. However, today is Sunday, and it is quiet and peaceful, with minimal traffic. Within seconds of leaving my house, a car paces up right behind me and honks. Despite the normally busy road, with this early hour and sparse weekend traffic, this car has two lanes in its direction. Meanwhile, the driver continues a short distance behind my rear wheel, honking every few seconds. I keep riding, doing my best to ignore them. "Beep… beep… beep!" continues, as I move over to within inches of the curb, to allow the driver to pass easily. Finally, after upwards of sixty seconds, the history of my harassment over the years takes the best of me, and I turn around with an aggressive gesture, challenging the driver to hit me. To my dismay, it is a nice-looking older woman, cringing and reeling back at my exclamation, like she saw some life changing destruction. I see her whole face rear in shock, like my frustration surprised her.

While her look of shock penetrated my zeal, I still felt victory, as she quickly drove around me. However, less than two minutes later, I come upon an older gentleman, riding his bike. He pedals slowly and wears black shorts, a black jersey, and a blue helmet. Suddenly, the violence of my act pierces my ignorance. Very likely, that woman was this gentleman's wife and kept patiently honking to get his loving attention. My prior sense of victory instantly turns to anguished loss, as the subjective nature of my reality shifts with new information.

See how my limited understanding shadows my perception of reality? So it is true with our view of the Divine. We must come to the task of understanding our Creator humbly, knowing our own limits. However, this should not dissuade us from seeking *It* for the understanding *It* provides.

C. Moral Code

In Part 1, I challenged you to consider both the science and philosophy of our existence, focusing on the physical world comprising our Universe. Now, I build upon earlier statements of moral reality, where the world and Universe are much more than their physical dimensions, or merely experienced from our five senses. There are realms and dimensions that extend beyond our tangible grasp, with an accompanying reality that is all-encompassing. I call these realities "moral" and "spiritual." Our physical reality, and the moral realities that accompany it, are a subset of these greater spiritual truths.

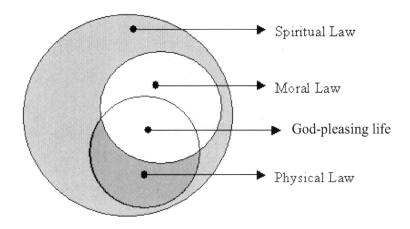

Whether we ascribe to moral law or not, we are still subject to truths that extend beyond our senses. While disastrous and frequently in direct conflict with governmental laws, we can exist outside of moral law. While we continually hear that physical law is the only law, it is the least important of the three.[74] The fact that the popular media seems to sing this tune in such harmony should only lend credence and support to the notion that we each must analyze, discuss, and bring meaning and understanding to a more objective perspective of our lives and our role in the Universe.

Moral laws are those behaviors and actions that humankind innately possesses, intrinsically conceived within each of us, largely when we are born. While they may be widely shaped and transformed by nature and nurtured with the world around us, they are the innate set of values we all have in the absence of culture and experience. The basis of this category comes from looking at the world beyond a limited-physical and three-dimensional lens. It delves into the notion that regardless of our time in

---

[74] Indeed, Jesus proclaims that if your eye tempts you to sin, it is better to gouge it than to let it send your soul into eternal hell. (Matthew 5:29)

history or geographic location and culture in the world, we all share a limited foundational set of views[75] of right and wrong.[76] Due to our actions (and those actions upon us), as well as our environment (and subsequent actions around us) that shape our perceptions of reality and social norms, moral law may change as we go through life. It encompasses both objective and subjective facets. However, just because our subjective views of moral law may change, it does not mean moral law changes. It includes absolutes, where the Universe may restore our original definitions of morality[77] and the inherent purity associated with our original state.

Example of Moral Laws

While by no means an exhaustive list, a few examples include:

1. Treat other people the way you wish they would treat you.

2. Do not murder.

---

[75] "This is the covenant I will make with the house of Israel after that time, declares the Lord. I will put my laws in their minds and write them on their hearts. I will be there God, and they will be my people. No longer will a man teach his neighbor, or a man his brother, saying, 'Know the Lord,' because they will all know me, from the least of them to the greatest" (Jeremiah 31:33-34).

[76] As in the time of Job, the first covenant dictated that the family head owned responsibility for communication with God and the moral well-being of his clan. With the introduction of the Levitical priesthood, the second covenant transferred this responsibility to a chosen Jewish tribe. This remained until the time of Christ, when not only is each person responsible for his own moral chemistry but able to commune directly with our Creator.

[77] "But you [followers of Christ] have been anointed by the Holy One, and you all have knowledge. I [Paul] write to you, not because you do not know the Truth, but because you know it, and because no lie is of the Truth… But the anointing that you received from Him abides in you, and you have no need that anyone should teach you" (1 John 2:20 and 27).

3. When one feels compelled to do the right thing, do it.[78]

4. Do not steal or cheat.

5. Do not give false witness against another person.

6. Do not covet or have ill-desire for property or possession held by another person.

Universal laws, moral realities of right and wrong, which lead to universal *Truths* exist. Many of my actions form due to these beliefs, frequently based upon more than what I want or desire. Emotions and feelings of the moment frequently contradict the heart and mind. Furthermore, each of us *know* some things to be *true* and *right* and others to be *false* or *wrong*. It is our subjective reality and our perception of reality. The things we say, coupled with the things we do, create a platform formed by our choices that carry us through life, interfacing into the world around us, shaped by our will and priorities. Similarly, our environment, both geographic and cultural, impacts this same platform, biasing our choices. Evidence supports notions of common moral reality. We see universal laws of objective morality, and while subjective views may change over time with daily experience colliding into cultural values, experience, and faith, we also see an inherent programming or moral mold from which we all start.

Two of my agnostic friends ask me how I can possibly justify my faith for supporting a certain political position. Their logic focuses on a request to take away federal funding for certain international humanitarian efforts. They see the moral need to support such causes, despite their agnostic

---

[78] "So whoever knows the right thing to do and fails to do it, for him it is sin" (James 4:17).

faith. In fact, we share morality on such issues, which is interesting that people who do not believe in a *higher being* feel themselves subject to moral responsibility. They admit to a tug on their hearts and consciences that implore them to help others along a set of shared values of helping the poor and those in need, and see such behavior as '*the right thing to do*'. It is more than instinct. Why would such moral responsibility have a place in a *survival-of-the fittest* and evolutionary world? The difference in view focuses simply upon responsibility. This example illustrates notions of shared morality, despite differing views on spirituality and of faith, and different ideas about how to express morality.

This desire to help others plays out in situations that might seem contrary to a purely survivalist nature, as well. One night, I am walking around town about one in the morning, when I hear a woman scream. Immediately, my heart starts pounding. A quick glance around the buildings, and I cannot see anything. Did I imagine it? Then, I hear what sounds like steps and muffled motion not too far from me. I rush forward, turn the corner, and see a man and a woman struggling, or fighting, with the man pushing the woman up against the brick of a building wall. I announce my presence, to which the guy glances at me with an annoyed and somewhat dull, yet confrontational expression. However, at the same time, the woman interjects that she is fine, thanks me, and says she has it under control. I watch for a moment, trying to ascertain whether someone needs help (not that I have any idea what I might do) until the guy yells some obscenities at me. The woman, again, shouts she has it under control. I slowly walk away, glancing back every few seconds to make sure the situation is not worse, or that I might get shot or attacked. While a purely survival response to such an encounter is to flee for self-preservation, I overcame this fear to stay. Most of us have these same strong, yet opposing responses pushing into our thoughts, actions, and emotions. The best evolutionary explanation is that I associate with the victim and feel a selfish desire to help, so that should I find myself in such a situation, someone might help me. However, this never even entered my psyche during my episode. Despite the inherent danger, I simply wanted to help. This is the case with most of us. Without morality, we interpret such a response as irrational, yet it further illustrates an innate programming and moral reality.

Humans oppose certain behavior, such as murder. While the end goals of protection, wealth, power, etc. may be positively received (and the foundation of most wars), humanity still sees such means as morally and socially inferior to competing efforts, like diplomacy. One set of people thought of as encouraging fierce and frequently deadly behavior as morally acceptable are the *Yanomamo* Indians in Amazonian rainforests of Venezuela and Brazil. An anthropologist who lived with and filmed them, Napoleon Chagnon, published books about how the men who were the best fighters received the most respect and status.[79] He uses the *Yanomamo* to illustrate how in their culture, they deem murder as good. Most cultures in history contrast this view, with overt violence deemed negative and bad. This might be considered a form of moral code (do not kill your neighbor) and is consistent throughout much of the world and through time. However, in the *Yanomamo* culture, due to shortages of resources (and women), this anthropologist argues, violence was deemed enviable, making

---

[79] Chagnon, Napoleon, <u>Yanomamö: The Fierce People</u>, 1968.

it contrary to the overarching moral code of society in other regions of the world and points in history. However, after watching some of this original video footage, simply in looking at the behavior of the other villagers, clearly, the clan fears the violent person. While having such a violent warrior might be advantageous to the clan's survival (and deemed beneficial), one could see in the behavior and interactions that everyone else simply gives respect primarily out of fear. Like a bully in school, the other villagers respond like students, trying to keep their distance from the bully, or violent tribal member, to ensure they do not erupt his wrath. Rather than refuting the common morality that murder is wrong, the *Yanomamo* example further supports it. In fact, today, many *Yanomamo* clans flee the jungle to escape the survivalist, murderous environment, despite their aversion to modern society.

Similarly, humans do not limit moral law to violence. Another interesting view with the *Yanomamo* that fits a common idea of ubiquitous moral law, is the notion that incest is bad. Despite severe limitations on the number and availability of fertile, age-bearing women, the notes and memoirs also paint the *Yanomamo* people as refraining from incest. While some may argue incest is simply a result of evolution, as a means of avoiding higher chances of genetic abnormalities, people must be consciously aware of such genetic liabilities. Additionally, even with such genetic liabilities, if purely based upon survival instinct, they must realize, in many cases, taking a woman from another tribe or village will mean war and possibly death; the low-risk option would then be to procreate with a relative to avoid such possible death. However, neither scenario is valid here. Even if the avoidance of genetic abnormalities is a subconscious evolutionary trait, without morality, incest would still be safer in most scenarios, due to the bloody retaliation with opposing *Yanomamo* tribes. Indeed, while the definitions of family relationship restricting such behavior vary, only in royal blood (and its obvious limitations), has it been considered moderately acceptable.

Another friend, Jason, believes human nature evolved into moral behavior. He uses this as a foundational argument to claim that there is no Designer or *god* in the Universe, simply random circumstance. This view intersects philosophy and biology, pulling on cultural norms of peer pressures and judgments in coordination with evolutionary behavior that shape psychological traits.[80] His logic follows evolutionary metaethics, focusing on evolutionary theories that "support a non-cognitivist account of the semantics of moral judgment... to undermine the claim that there are objective moral values..."[81] In this view, morality is simply a construct of evolutionary influences that shape our thoughts and feelings on certain issues we consider moral.[82] Rather than discounting such views, I simply argue that the construct of moral values not only fails in dismissing a godly Designer, but supports notions of Deity. How man received such innate moral programming simply illustrates the great lengths academic people go

---

[80] Fitzpatrick, William, *Morality and Evolutionary Biology*, Stanford Encyclopedia of Philosophy, 2014. Accessed December 12, 2017 via https://plato.stanford.edu/entries/morality-biology/.

[81] Fitzpatrick, William, *Morality and Evolutionary Biology*, Stanford Encyclopedia of Philosophy, 2014. Accessed on 12/20/2017 at https://plato.stanford.edu/entries/morality-biology/.

[82] Street, Sharon., *A Darwinian Dilemma for Realist Theories of Value*, Philosophical Studies, 2006. Volume 127, pages 109–66. Accessed on 2/1/2018 at http://fewd.univie.ac.at/fileadmin/user_upload/inst_ethik_wiss_dialog/Street__Sharon_2006._A_Darwinian_Dilemma_for_Realist_Theories_of_Value.pdf

to discount a Creator and foster their own cold and gloomy self-deity. Such intellectualism expands subjective realities to become the focus of their energy and even *worship*. They miss the beauty of Creation. What is important, and what they acknowledge as true, is that we all have moral values and that these values are even shared across people groups, geography, and time. While the *means* claim significance, it requires too much dry discourse and really just detracts from focusing on the *end*.

A desire to help those in need, while abstaining from murder and incest appear to be common facets of moral law and examples of objective reality. Numerous examples throughout time and region conform to this morality, with examples contradicting it being few and easily refuted, even with cultures completely cutoff from the outside world, like the *Yanomamo*. There is a consistent theme in these perceptions across humanity showing an objective reality. Shared morality supports notions of spiritual truth and universal deity.

Humans also share positive moral values. Notions of love, for example, support the pre-disposition of moral character. Do you love someone unconditionally? While other languages have multiple words for love, there is just one word in English. In Greek, for example, it leverages three words, *phileo*, *eros*, and *agape*. Simply defined, *phileo* refers to brotherly love and the love you might have for family and friends. *Eros* refers to the passionate and sometimes physical love between a man and a woman. Meanwhile, *agape* references the love between God and humankind, such as a small glimpse of the selfless and sacrificial love between a parent and a child, or lifelong trusted friends. Ideas supporting Darwin's theories [83] of *survival of the fittest* and self-preservation associated with *natural selection* make sense with *philēo* and *eros* and bring credence to evolutionary love.

However, this is not the case with *agape* love. It is an overpowering love that transcends friendship and passion, driving into the heart for pure action that is typically self-sacrificing. It is love that does not care about self-preservation, self-promotion, or any type of personal aggrandizement, seemingly counterproductive to Darwin's impersonal agnostic evolution. A mother that gives her last meal to her child shows a certain form of instinct for survival of the family. However, if the child is too young to guarantee his/her own survival, the survival instinct from the mother would be to let her child die. Across history and societies, this happens, but most would consider it a rare anomaly, rather than the norm. If not from a Designer or Creator, *agape* love would be viewed as a negative trait and weeded out of the evolutionary line thousands of years ago. *Agape* love holds tremendous power and does not fit arguments for a world without a universal Creator. It is a simple proof that reflects personal character in the embodiment of this *Being*, offering love for us expanding upon pure evolutionary need.

Humanity shares a consistent theme in multiple aspects of morality, which illustrates an objective reality forming these threads of ethical norms. In other words, morality is not merely a by-product of physical reality, but rather the foundation of this reality. Moreover, as with the innate order of our physical Universe, the foundation of moral reality suggests a designed heart and mind. Shared morality suggests notions of spiritual truth and a

---

[83] Darwin, Charles, <u>On the Origin of Species</u>, 1859, Harvard University Press, published 2001.

moral, universal Designer who formed us. Furthermore, it directs us to a clearer understanding of the underlying spiritual reality on which it depends, which we explore in the next section.

D. Spiritual Law

Spiritual laws, on the other hand, presuppose moral and physical laws. Moral laws exist due to, and as a by-product of, spiritual law. Rather than drawing upon limited physical science to qualify their viability, we see spiritual law from seeking perspective on deep spiritual rationale and intrinsically gleaning insights about moral reality. They tie into the presumption of an *open-system*, existing outside our Universe, supporting notions of a greater reality than that which man readily sees in typical daily pursuits. In leveraging spiritual laws, we reflect upon the way the world must work, making projections about subjective perceptions of reality and our Universe. They reflect more sublime truths about our lives, actions, and our environment. They ask the question 'why' there must be moral laws. What is it about our world and the way in which we relate in the Universe that causes the inherent truths surrounding moral law? It questions why words and deeds that separate us from our Creator become wages we earn, which result in the payment of death.[84] Spiritual laws govern how foundational truths interact with humanity and other forces within the universal spiritual realm.

---

[84] Sin may be defined as a turning of man's heart through freewill away from God and to ourselves, with our own self-reliance and self-deification, so to speak, being the root of such sin. When we turn away from God, we disrupt our ability to fellowship with Him. In not seeking God, we live out of our own strength, losing out on the tremendous value we get from continuous interaction with the Creator of the Universe.

Like an apple falling from a tree and our belief in gravity, spiritual laws affect us, whether we believe in them or not. With an *open-system*, the door opens to spiritual realities. Many of them make little or no sense to me, yet my lack of appreciation or understanding does not nullify their truth or implications. For example, there is a Judeo-Christian-Islamic belief that actions which take us away from honoring the God that created the Universe demand a blood penalty. While I may not understand or even agree with it, spiritual law still binds me with its implications and consequences. Inherent truths associated with this plane of reality place demands on the Universe and each person in humanity. These laws are characterized in absolute Truth, as they supersede both relative physical laws and moral laws.

Examples of Spiritual Law

While I jump the gun by listing some spiritual laws biased by the *Bible*, they still hold as examples falling under the definition of spiritual law. It is by no means an exhaustive list, but a few examples include:

1. The things we say carry great consequence, to bless or curse, and we must manage our words carefully.[85]

2. We, man and woman, are made for relationship, first with our Creator, second in community with one another.

3. All actions not based upon *faith in* and *honor of* the *Force* that created the Universe separate man from

---

[85] "And if anyone does not stumble in what he says, he is a perfect man, able to also bridle his whole body" (James 3:2). "The tongue is set among our members, staining the whole body, setting on fire the entire course of life, and set on fire by hell" (James 3:6).

relationship with *It* (ie: sin).[86]

4. The desires of moral reality fight against the flesh and the desires of the flesh fight against the spirit.[87]

5. Actions not honoring God demand a blood penalty, or sacrifice, to which we all must succumb.[88]

6. Separation of relationship with the universal Creator brings physical and spiritual death.[89] A consistently moral life equates life in perfect coordination with God.

7. There are only two ways to stand before God: the Law or a perfect sacrifice.[90]

8. This is impossible to do as a mere mortal, as all have sinned and fall short of the glory of God.[91]

9. The need for a sacrifice expands to include all humanity; it requires a life without sin or separation from God.[92]

---

[86] "Everything that does not come from faith is sin" (Romans 14:23).

[87] "'For the sinful nature desires what is contrary to the Spirit, and the Spirit what is contrary to the sinful nature.' They are in conflict with each other, so that you do not know what you want" (Galatians 5:17).

[88] "I am the way and the truth and the life. No one comes to the Father except through Me" (John 14:6). "But God demonstrates His own love for us in this: While we were still sinners, Christ died for us" (Romans 5:8).

[89] "For the wages of sin is death, but the gift of God is eternal life in Christ Jesus our Lord" (Romans 6:23).

[90] "All who sin apart from the law will also perish apart from the law, and all who sin under the law will be judged by the law" (Romans 2:12). 'Law' refers to the Moral Law of the first five books of the Bible (Torah) and Jewish canonical doctrine.

[91] "For all have sinned and fall short of the glory of God, and are justified freely by His grace through the redemption that came by Jesus Christ" (Romans 3:23-24).

[92] "I tell you the truth, whoever hears My word and believes Him who sent Me has eternal life and will not be condemned; he has crossed over from death to life" (John 5:24).

E. Losing One's Life to Find It

A simple and reproducible proof supporting spiritual reality looks at seemingly mundane acts of service. Rather than trying to find one's life, this spiritual proof focuses on trying to proactively "lose one's life" by performing activities one may already be doing, yet with a focus on doing them to honor our Creator. When we proactively lose our life for this *Being*, even through every day activities, we find peace, joy, and purpose. With our culture of self-gratification and personal entitlement, it is counter-intuitive that as we seek to create the life we want, we seem to never find it. Meanwhile, when we focus our eyes on blessing those around us in the name of the God-Most High,[93] we may not do what we think we want, but we get the life we crave and need.

This seems to be even more relevant in today's culture, which looks to a variety of cures for lack of peace, joy, and purpose. Although many real world issues can benefit from chemical cures and drugs, or even exercise, we tend to make medication the first choice, rather than one of the last. We have become an overdosed culture, treating a host of bouts of despair, missed dreams, and expectations. Despite a world where technology enables more ways of communication and sharing than any time in history, with an ease of life not even attained by kings throughout history, doubt, despair, and depression plagues a large portion of the population. People have a vain expectation that happiness is a state of finality, or a finish-line. We oftentimes hold high expectations and

---

[93] In reference to King Melchizadek, to whom the newly created Jewish nation, through Abraham, offers tithe and blessing, "'And blessed be God Most High, who delivered your enemies you're your hand.'" (Genesis 14:20).

overflowing feelings of entitlement that keep us from the contentment we seek. In today's culture, we mistakenly seek to mold our lives and work into posts for social media, comprised of pristine statuses and perfect pictures. We then feel amazed and bewildered when we feel alone and empty. Our Creator did not form us to find and create our own lives apart from Him.[94]

For two weeks over Christmas vacation, I did something I genuinely despise doing, simply to test this proof. While visiting my family, I claimed the right to do the dishes for every meal. At first, I admit, my self-absorbent focus was on how I hate doing the dishes, but as I shifted my thinking and focus off myself and began to use the time simply as an unspoken means of loving my family and honoring our Creator, I began to feel almost *holy* about it, practically craving it. I felt blessed and even (for those that know me, do not laugh) perhaps, angelic. The more I gave, the more I received, even feeling blessed for something as simple and trite as doing dishes. I am not sure anyone in my family even noticed, as I received no words of appreciation, yet I still felt warmth and love flow into me. Even years later, I still make a habit of doing the dishes over family holidays.

Another time, I tried to think of how I could really lose my life to bless someone else. I heard someone mention that they need volunteers to teach at a high school for students that get kicked out of the public school system for drugs, gangs, violence, or pregnancy. I thought my boss at work would never let me take time off during the day, but he said if I

---

[94] "For whoever wants to save their life will lose it, but whoever loses their life for Me will save it" (Luke 9:24).

stayed committed to the completion my projects and worked outside normal hours to meet demands, he was fine with me taking a few hours each week to volunteer. Again, at first, I focused on my needs not being met and almost quit, but after meeting with the kids a few hours every Monday morning, I slowly began to view it simply as a way of losing my life for our Creator, seeking to give 100% of myself with nothing in return. I taught grammar, spelling, reading, math, and even typing. Again, the more I gave, the more blessed I felt. I remember many days, leaving my class to head back to work, thinking that I got more out of volunteering, than I did with a job I thought I loved. In pouring myself out, just by offering my few skills to a couple dozen students, I felt overwhelmingly blessed. This did not last forever, as I began traveling for my job and could not keep the original schedule, and then I ended up leaving my job to move to another state. Nevertheless, the proof holds true.

Once more, I decided to volunteer at a homeless shelter. I am pretty goal-focused, so I like to *do* things. I thought that if I was going to volunteer for a few hours, I wanted to be productive. However, their only need was for me to staff the front desk. Unfortunately, almost everyone there already knew what to do, so I really did nothing. I just sat there, hour after hour. No one even asked me a question. I felt it was a total and complete waste of my time. After one day a week for a few weeks, I, again, almost quit. I had to transform my mind from being 'productive' to just 'being'. I started bringing my *Bible*, and used it as a time of reading, reflection, and praise. I slowly learned the names of some of the frequent guests, and began talking to them. As weeks turned to months, people got used to me and thought of me as 'the guy that always reads his *Bible*'.

They began to ask me questions about my faith, then about work and about life. I began talking with more people and got more involved in their lives. I slowly *found* my life, as I *gave it away*;. Unfortunately, this, too, did not last, as I changed jobs again, but repeating the spiritual proof resulted in blessing and peace.

If not already, try it yourself. For many of us, we are so used to trying to *find* our lives that we may need to look for opportunities to *lose* them. What do you hate most that may very well be a service, or blessing, to others? Perhaps, look for ways to do that one thing. As we give of ourselves, we shift our focus off our needs, our timelines, and our expectations, removing them from our minds, words, and actions. In return, as we replace such selfish pursuits with a 100% focus on how our Creator wants to use us, with the intent on only giving, we find ourselves. Almost miraculously, yet as in a reproducible, scientific method, when we seek to lose 100% of ourselves for God, frequently in service of others, we find peace and purpose; joy springs up within us that transcends our understanding. If not already, try this experiment and lose your life as a form of reverent worship.[95]

Giving genuine praise to our Creator results in the presence of God pouring into and around us. I feel God most consistently when I reach out to *It* through proactive thoughts and words, thanking *It* for some of the blessings in my life. This can be spoken aloud or just in my mind, and it may focus on any variety of topics. I like to do this while riding my bike and driving in the car, but it could be anywhere. Although I frequently

---

[95] As many people get stuck on ways to lose their lives, although many means are possible, one resource for volunteering: http://www.universalgiving.org/volunteer/

praise God simply for my *being*, it could be any facet of health to a friendship or conversation, including the taste of a favorite food, sip of a favorite drink, sight of a sunrise or sunset, a sensation from any one of our senses, or the fact that my car happens to be working at that point in time. All of us can find dozens, if not hundreds, of things for which we are grateful, if we think about it. To validate this proof, we focus upon our Creator as the source of our gratitude.

Some days, every once-in-a-while, I feel so depressed that I do not want to get out of bed. Work may not be going well, or I feel my body stressed and run down, as fighting off the flu or illness. I may not feel God and cannot remember recently feeling close to Him. It sends me into a downward spiral. While driving to work, I simply start singing to our Creator, laying bare my situation. Slowly, I begin to start proclaiming ways I feel blessed, and my heart shifts from me to *It*. In just minutes, I feel awash in the presence and love of God. It fills my soul, then begins to overflow. Frequently, I must fight back the tears of praise that seek expression. It feels so good. This is typical; I do not understand how people make it through life without this regenerative relationship with our Creator. God wants it for me, and *It* wants it for you.

F. Miracles

Many people feel that while miracles may have occurred in the past, they do not transpire today. However, they most definitely do. They happen in my life repeatedly, from the simple to the far-fetched, and they can happen in the lives of all people that know the Spirit of God. They illustrate a wonderful proof in how spiritual realities intersect physical reality. While I typically see God respond consistently with nature, it is usually far from natural.[96] While people discern God in a wide variety of ways, for me, a strong yet strange idea usually pops into my head that oftentimes, I do not want to do, yet have a feeling of peace about doing. However, others may see God in nature, through sleeping dreams, waking visions, reading, or by one or more other people, etc. There are many other ways for God to speak to us, too, so it becomes more important to filter out all the background noise that clouds God's soft voice, to slow down, pause our busy schedule, and devote time to listen. I relate a few stories experienced and seen that may bring meaning to you about how God cares for us and even knows our every intimate detail.

1. Passenger

A few years ago, I was passing by the San Jose, California airport (SJC) to cut over to interstate 880 south. It was a hot summer day, being humid and almost 100 degrees Fahrenheit. I had been fasting and felt a closeness with God, although at this particular moment, my heart felt heavy, and I cried out that our Creator might use me in some way, great or small. Instantly, a strong, yet somewhat random idea popped into my head to stop the car. This is pretty easy to do, so I thought, "Why not?" and did it. Once I pulled over, I looked in my rearview mirror and realized I just passed a man walking down the sidewalk, pulling a piece of luggage. Not really sure what to do, I rolled down the window. As it

---

[96] While I limit my examples to plausible, yet unlikely, events, God also acts outside of nature; I have examples of events outside our definitions of science, too (yet do not include them herein).

took a good twenty to thirty seconds for the man to approach, I wondered whether I was crazy.

His rolling suitcase squealed as he walked, and once close, I leaned over to the passenger window to ask him a question. "Can I give you a ride somewhere?" I asked innocently; I had no idea what to expect. Amazingly, he responded, "Sure!" He opened the car door, as what was left of the cold air in my air-conditioned car swept away, replaced by the hot, moist air outside. "I am just heading to the private airport," he offered, which was only about a half mile away, and he jumped in the front seat. I made some comment about the hot weather and how the car air conditioner would cool things off again in a minute, as I already felt my body begin to sweat. There was that awkward pause, as we both pondered what to say next. Then he startled me, offering, "Are you, by chance, a Christian?" To which I hesitantly responded, "Why 'Yes' I am, why?" He then ventured, "Well, you may think I am crazy, but not 30-seconds before you stopped, I prayed to God that if He really loves me, that someone would give me a ride to the terminal." I laughed aloud and responded, "That is hilarious, as less than 30-seconds before I stopped, I prayed to God that He might use me, no matter how simple or grand." By this time, we were already at the terminal, so he hopped out, and we kind of looked at each other with a fellowship of amazement and smiled, and he just walked away. I drove away praising the universal Creator.

2. Hitchhiker

The *Force* of Creation may also use seemingly random circumstances toward a profound orchestration of events. Again, following a time of fasting and prayer and seeking His will to use me in the upcoming year, I plan to drive from Boulder, Colorado to Redlands, California. It is early April, and I have made this trip many times before. Just as I get ready to leave, a strange thought jumps into my mind; I feel compelled to pick up a hitchhiker. Several years prior, a buddy and I hitchhiked across different parts of the world. Upon my return, I picked up almost everyone I saw. However, after a series of bad experiences, I kind of just stopped. It is not really a formal decision, but I just find myself no longer picking up people along the roadways. Thus, this sudden thought popping into my head seems a bit out of place. Still, I accept it, almost as a cause for adventure, and as I prepare to leave, I look forward to meeting the person I will pick up.

I finally jump in the car and head out of town. Just as I hit 93 south, on the fringe of Boulder, a hitchhiker pops up on the edge of South Boulder Road. I ponder whether this is my passenger, but immediately hear an idea take immediate shape in my head, "Nope, not this one." I see several more hitchhikers along the way, and every time, it is the same thing, "Nope, not this one." Finally, after several hours of continuous driving, I get all the way to Las Vegas and pull into a small casino for their all-you-can-eat buffet. I have less than $30 (and a debt-ridden credit card) to my name, so I try to maximize my budget by capitalizing on the 'all-you-can-eat' aspect of the buffet.

Upon walking into the casino, I discover the buffet line has a twenty minute wait. While I do not condone gambling, I played a few hands of $5 blackjack. I won $80, almost tripling my net worth. I stroll back to the buffet line, now feeling good about paying for the meal, but they tell me they already called my group, and they want me to wait another twenty minutes. Fearing losses with a return to the blackjack table, in frustration, I walk out of the casino hungry.

It is early evening, with the sun already disappearing behind the horizon, yet the sky offers a soft, flat pinkish hue; it is dark enough to make it hard to see, but still where headlights do not really provide much value. I begin circling the onramp leading onto the freeway, still thinking about the fact I have yet to eat, when a strong thought interrupts my selfish angst about missing my buffet, "Stop! This is the hitchhiker!" I do not see anyone, so I keep going for a split second, but the thought gets really loud, "Stop now!" I pull over, although more out of curiosity, than anything else. I do not really know if God is speaking to me, but decide quickly that it does not hurt to validate the idea. I turn around and see some dark shape running toward me along the shoulder, about 100 meters back. I back up the car, until we meet.

The figure amidst the fleeting shadows is a guy hitchhiking to San Francisco. If one looks at a map, Barstow is kind of an intersection point between my route to Redlands and Vegas to San Francisco, so I offer to take him to Barstow.

The hitchhiker, call him James, pulls open the door and settles into the front seat. He has no luggage of any kind and is just wearing jeans and a t-shirt, despite the cooler evening air. Once settled, we start talking. While I have no idea what is true, he shares that he is eighteen and ran away from home about two years ago. He looks older, but there is really no point in questioning it. He reveals that he grew up in Half Moon Bay, where he lived with his mother. However, his girlfriend (call her Veronica) at the time moved to Las Vegas to become a dancer, and James followed her a few months later. James had not spoken to his mom since he ran away from home, but Veronica speaks with her on his behalf every few months. When he first came out, both he and Veronica got sucked into the sex/drugs culture in Vegas and developed their chemical addictions.

James longs to return home. Every time he tries to leave, however, someone in his and his girlfriend's counter-culture sees him and brings him back to his apartment in Las Vegas. He remains dependent upon them through his chemical addition. After over a year of this, he feels desperate about returning to his family and a balanced life, back to the safety of his mother. He shares how just an hour earlier, he got down on his knees and cried out to God amidst the darkness around him. At this moment, while on his knees, he suddenly feels compelled to leave, rushing out of the house without money or even a jacket. This all happens just minutes before I see him. He says he had only been standing there a few moments when I stopped. He did not even have time to stick up his thumb, as is common for hitchhikers.

We continue talking about our lives, with me sharing my faith in our Creator. He shares that his mother is a Christian, which he resented at the time, and while James knows about God, intellectually, he does not yet interact with our Creator on a personal level. He desires to know God, though.

As we pull into Barstow, my strategy is to drop him off at the bus station or a location where James might find other travelers and people to continue his hitchhiking. By this time, it is after 11pm, and I do not realize how I am more focused on the conversation, than my driving. We continue to talk, until realization hits me that I am totally lost. I have been driving around aimlessly, so I pull into a gas station to ask for directions on how to find the bus station. The attendant replies that although the bus station is back on the other side of town, the train station is just down the hill. I figure 'why not' and drive to the train station.

As we pull into the station, I cannot see any other people. It is quiet and empty. All the lights are off, but as we pull in, they all turn on, like they are on a motion detector or something. I jump out of the car to search for a schedule when what looks like a young couple with two kids runs up to me. They begin yammering about various topics and are all jumpy and emotional, asking me where I am going. I share my desire for a possible train to San Francisco, or somewhere that general direction, at which their energy level increases even further. They are practically bouncing, they are so full of energy. The young man keeps talking the whole time, and I am not able to keep up with all that he says. I have no context to what seems like a complex situation. Apparently, he has a ticket to Seattle, the train goes through the Bay Area, and he is desperate to sell it. I offer him $60 of my $80 black jack winnings, to which he immediately accepts, yelling, "Praise God!" By this time, this family all begins laughing and crying, and jumping up and down. Through all their tears and laughing, he relays some of his story.

The young man with the train ticket mentions something about wanting to stay in town, but not having the money to buy another fare. I think he is in the military. The travel must be that day, and they spent the entire day at the train station trying to find someone to buy their ticket. The last train was in just a few minutes, but they had been waiting for several hours since the previous train and were about to give up. They walked across the street to sit in a field, praying for a miracle for hours. He so desperately wanted to stay a few more days with, it seemed, his family, but had commitments in Seattle and was on a tight budget. Finally, they saw me pull into the empty parking lot and ran over to see if I might buy their ticket. They did not even have the chance to offer it to me before I asked if they might sell it.

I exchange the money for the ticket and walk back to my car, ticket in hand. Meanwhile, James has been sitting in the car the entire time, wondering why the family is yelling and jumping all around, hugging one another and me. I hand him the ticket, tell him that God loves him, and I hope he makes it home to his mom. I give him a box of energy bars and an old jean jacket. With this gesture, his very reserved exterior breaks down, and he begins to tear, overwhelmed with God's love, thanking me profusely. He walks into the train station (which is now, timely, open). I tell the family that I, too, try to follow God, at which they start crying, laughing, and hugging all over again. I walk to my car, and drive away. My soul cannot help but praise a living Creator. What a profound orchestration of people and events. I try to listen to music, but I do not hear the words. The remainder of my drive to Redlands, I simply sit in awe, goosebumps hitting me from time-to-time. I no longer feel my hunger. I laugh, loving the irony that God just used me (of all people). I feel simply overwhelmed and awash in peace and love.

3. Healing

My father is now about sixty-seven-years old and cannot maintain energy. However, he has yet to find appropriate respect for the medical field, and he rarely visits a doctor. Finally, when he begins coughing up blood, his wife, my mother, makes him see a physician. He goes but keeps the results from her. However, he calls me to confide that the doctor says the cancer advanced through his lung and gives him three months to live. He is not one for drama and simply tells my mom that he was diagnosed with lung cancer, but nothing about his doctor's pronouncement that it will be a miracle if he makes it through Christmas to the new year, just three months away. The cancer is so advanced that the doctors question whether to even attempt treatment.

My dad has the constitution of an ox, however, faring better than the doctors expect. After a few weeks, they decide to try radiation and chemotherapy. This continues and after nearly three months of both treatments, more tests show that the cancer is spreading and worsening. Nonetheless, the doctor needs more tests to be sure. My dad calls me again, sharing the doctor's prognosis. In anguish, I begin fasting from all food and the next day, spend a few hours in prayer.

I feel what seems like God placing ideas in my head. I see a very clear image in my mind of what can best be described as fused strands of tissue, not unlike an extremely unruly, knotted, head of hair. In my quiet prayer and meditation to our Creator, I begin proclaiming that the chunks separate, as if someone combs out the fused strands of hair. As one places a comb through severely knotted hair, some hairs get yanked and ripped out. So it was in this waking-prayer vision, with cells and fibers tearing. After an unknown period (maybe an hour or two), I stop, feeling a sense of peace about it. Not being sure what this means, I call my dad just to share that I think his health will improve and that he might be in pain for a week or so. He refuses to understand my statements, but accepts my call at face value.

A week later, he travels back in to the doctor for follow-up tests. To everyone's surprise, the doctor shares that what he thought was a continued proliferation of cancer is, in fact, scar tissue. The physician exclaims that it looks like all the cancer cells have been surgically removed, leaving his lung filled with scar tissue in its wake. My dad subsequently recovers and lives another seven years before finally succumbing to side effects related to treatment on other forms of cancer.

4. Car[97]

It is a sunny summer afternoon, with a few fluffy white clouds amidst the hazy blue sky, as I drive to meet friends in Atlanta, Georgia. I head south on 400 (a north-south highway), when I *hear* or *feel* a voice pose a strange idea in my mind. While not audible, I interpret it like a voice, similar to someone speaking to me. I understand the white car is going to hit me. My first thought is, "How bizarre!" I look around, somewhat bemused, and see only one *white car*. In my rearview mirror, I see a white car about 100 meters behind me, moving up fast in the far left lane, whereas I am in the far right lane. Curious at the validity of the strong thought, I keep watching, almost as if for entertainment. The car moves over to the middle lane, as it comes upon a couple cars blocking it in the left lane.

This white car is probably going about 15 mph faster than traffic. Just

---

[97] Driving story occurred August 13, 2016.

as it starts coming up on another car in the middle line, it passes me, as I am still in the far right lane. At this particular section of 400, there is an exit up to a ramp at Lenox Road. With the off-ramp, the shoulder narrows, with a big concrete wall growing with the height of the off-ramp to where Lenox crosses over 400. Meanwhile, the car in the middle lane is going much slower than both of us, and this white car is not able to get far enough past me before it comes up right onto the slower car's bumper. I am just one lane over, almost behind the white car's right rear bumper, also sometimes known as the "blind spot". Rather than slamming on its brakes, the white car suddenly swerves into my lane. Because I have been watching it this entire time, I am keenly aware of everything around me. Instantly, I slow down and swerve into the what is about half a lane of shoulder, inches from the concrete wall for the off-ramp, which is about 15' high at this point. I avoided a collision from the white car and into the wall with mere inches of separation, yet without issue. Just as this whole thing happens, the driver of the white car realizes I am next to him with nowhere to go, and he swerves back toward the middle lane, only to realize the car he is passing is still there. As a result, that car starts to swerve into the far left lane, just ahead of the other two cars the white car passed previously. Again, the white car realizes he has nowhere to go, so he jumps back into my lane, fully committing to the full lane to get around the car partially swerving left out of the middle lane. However, by this time, I already slow down enough for the white car to squeeze just ahead of me. He pops right in front of my bumper. This entire series of events takes less than two seconds. No worries, despite several cars swerving across the road, we are all safe.

Sometimes, I look at my phone when I drive, so under normal conditions, it is likely that this car would have taken out the front of my car at 75+ mph, and a good chance I would have flipped or flopped into the wall, also likely getting very injured, if not dying. Thanks "voice"!

Miracles happen today. They happen to me regularly, they happen every day to people around you,[98] and they can happen to/with you. When miracles occur, we encounter God. Spiritual reality becomes real. They could be stories of healing and timing like those I share, prophetic words of truth, or insights of a profound or common nature. In looking at miracles, we focus attention to the supernatural and highlight God's desire to commune with us, allowing us to see a spiritual reality. Miracles point us to God. The closer we draw to this *Entity*, the less crazy it seems and the more real it becomes. What is most amazing about these four stories is that this personified *Force* uses me. There is nothing special about me. Anyone who knows me is probably more apt to share one of my many imperfections over any type of 'holiness,' and I will be the first to admit that I am a definite work-in-progress. My only possible novelty is that through many other stories I do not share here, I have the faith to know that He can do all these things and more. When we align ourselves with God and His will, they tend to just happen.

---

[98] For a powerful set of better qualified miracles, read <u>Defining Moments</u>, by Bill Johnson, who shares well-known miracles of bringing the dead to life and dramatic transformation. Johnson, Bill, <u>Defining Moments</u>, Whitaker House, 2016.

*　　*　　*

An intelligent *Force* created the Universe out of nothing but *Its*[99] nature. It is amazing that such an all-powerful *Being* offers us, humanity, opportunity to see and pursue *It*. The Creator of the Universe is, by definition, God.[100] We just walked through a series of spiritual proofs shining light on forms of global morality, with moral laws emerging as the result of spiritual truths. These examples illustrate a moral nature and values, which are intrinsic to mankind. We hold a pre-disposition toward supporting these moral codes and actions. Our beings are limited to our senses and dimensional limitations, yet there is more than just physical reality, with moral and spiritual realities extending through each of us, even uniting us. It is with and through the power and design in Creation that we can hope to step above worldly limitations and truly see absolute, or objective reality. There is more to life, and we, as humanity, simply need to observe, seek, see, and listen to this universal Creator. He offers us adventure. Like a Mobius loop or the fabric of physical reality, our efforts, albeit via the scientific method or a variety of other means of determining objective reality, we get to travel the path set before us, accept our limitations and strengths, acknowledge, and if we make such choice, seek to know this physical and moral Creator.

In this chapter, I established that evidence suggests both moral and

---

[99] Although the Bible states in Genesis 1:27 that both men and woman are "Man", using the masculine form of humanity for both sexes (male and female).

[100] *god*. Dictionary.com Unabridged. Random House, Inc.
http://www.dictionary.com/browse/god (accessed: August 24, 2017). "Noun, 1. the one Supreme Being, the creator and ruler of the Universe."

spiritual realities, with a multi-dimensional God who created our physical reality. In the next chapter, I provide a three-pronged view of the *Bible*, looking at whether it provides an accurate perception of science, history, and prophecy, enabling it to be a credible source for sharing further moral and spiritual truths.

Part III

The Bible

The Bible

In Part One, we established that the Universe is an open or non-isolated system with complexity and character, suggesting intelligent and purposeful design. In Part Two, we saw further credence to this design as God, with both moral and spiritual realities that impact our physical existence. In this chapter, we delve into why the *Bible*, unlike any other book, presents consistent physical, moral, and spiritual truths and represents a valid interpretation of reality. I will argue the validity of the *Bible*, looking at it from three perspectives: scientific, historical, and prophetic accuracy.

Starting with a definition, *religion* is a system of beliefs or ideologies defining the divine, incorporating specialized or ceremonial conduct and action.[101] A plethora of world belief systems meet this definition, from Buddhism to Sikhism, or ancient Greek mythos to the subjective belief systems and New Ageism.

---

[101] *Religion*, "An expression or belief in conduct or ritual of a divine or superhuman power or powers to be obeyed and worshiped as the creator(s) of the universe; any system of belief, worship, conduct, etc, often involving a code of ethics and philosophy; the state or way of life of a person in a monastic order or community", Webster Dictionary.
Dictionary.com - *n.*. (1) a. Belief in and reverence for a supernatural power or powers regarded as creator and governor of the universe. b. A personal or institutionalized system grounded in such belief and worship. (2) The life or condition of a person in a religious order. (3) A set of beliefs, values, and practices based on the teachings of a spiritual leader. (4) A cause, principle, or activity pursued with zeal or conscientious devotion.

Looking across the main nine religions,[102] scientists leverage what we know about physical reality as a measuring stick for the accuracy of each. Rather than delving into the analysis of these belief systems, several books already address the scientific plausibility of each of these world religions.[103] Out of these nine, I focus on Judaism, Islam, and Christianity. Other beliefs, such as Jainism, uphold many similar views, such as the need to tell the truth, avoid violence, help others, and hold to high moral integrity. Additionally, Jainism promotes a physical denial of certain passions like greed, lust, and pride, yet it is more of a philosophy of how one should live. I choose these three for their monotheistic beliefs in a single universal Creator. Meanwhile, Sikhism is also monotheistic and shares many teachings from these three faiths indirectly, as well as worship the Creator God, yet Sikhs confide they do not know this God nor do they ascribe notability to any part of the *Bible*. Rather than repeat this analysis, I recommend those interested to research each directly.

One of the most basic tenets of Christianity, Judaism, and Islam is in the commonalities across each respective canon, or book of teachings. All three share a common heritage and acceptance of the Torah[104] being inspired by God. Of the world's population of approximately 7.5 billion people, around 2.2 billion identify with Christianity, 1.6 billion with Islam,

---

[102] Christianity, Islam, Hinduism, Judaism, Buddhism, Sikhism, Shinto, Jainism, and Taoism

[103] Dr. Hugh Ross is a Canadian astrophysicist that was once agnostic and sought to disprove religion through refutation of the science described by the associated religious teachings. He was successful for all but the three Jewish-derived faiths of Judaism, Islam, and Christianity, with his analysis published in The Fingerprint of God and other works.

[104] Also known as the Taurat or Tawrah in Arabic for the Islamic faith; Quran, 5:44, or sura 5 (Al-Ma'ida), ayat 44

and 14 million with Judaism.[105] That is over 54% of the world's population. Just for the impact on civilization, everyone should read this shared framework. Additionally, it contains intricate detail on science, explains moral reality, depicts spiritual reality, references tremendous historical detail, and offers many levels of prophecy. This multifaceted value continues and expands in what Christians have as the canon of the *Bible* across both the Old and New Testaments and the Qur'an for Muslims. Throughout the Qur'an, Mohammad supports not only the God of the Jews (which is shared by the Christians), but its prophets,[106] including Jesus, John, and Jesus' disciples.[107] In the rest of this chapter, I will present a few of the many ways in which the *Bible* lives up to its claim to be a trustworthy record for not only physical reality, but also moral and spiritual realities.

Some people believe that small inconsistencies with scientific thought or historical consensus disproves the *Bible* altogether. This idea looks at where this shared writing may open itself to scientific vulnerability. If it is to be the word of an all-knowing God, simply recorded by man, it must be scientifically infallible. If it is merely a written history of men, not only is it fallible, but it serves as little supernatural value. However, rather than considering what scientists and historians think today and their subjective theories about the physical Universe and world history, it must be

---

[105] *The Global Religious Landscape*, The Pew Forum, The Pew Research Center, Washington DC, December 2012. To view the online version of this report, visit: http://www.pewforum.org/global-religious-landscape.aspx.
[106] "Muslims consider themselves sons of Israel, Abraham, and Isaac, same as the Jews." The Qur'an, Al Baqarah, 2:122.
[107] "Indeed, those who believed and those who were Jews or Christians or Sabeans those who believed in God and the Last Day and did righteousness will have their reward with their Lord, and no fear will there be concerning them, nor will they grieve." The Qur'an, Al Baqarah, 2:62.

compared to absolute proofs. History tells us that despite modern science and the brilliance of modern civilization, scholarly theories and seeming proofs are not always correct. Oftentimes, what is certain to one generation is simply naïve to the next. As most scientists will tell you, discovery merely brings more questions and more unknowns. In this line of logic, just because science describes a phenomenon, does not mean that we understand it. Take gravity, for example; it is a force that almost everyone says they understand. However, just because people observe it and can describe its affect, does not mean they know the reason how and why a mass provides gravitational attraction.

As one delves into the *Bible*, one sees it is clearly very different than other books throughout time. The *Bible* is the collective work of dozens of authors telling a cohesive story that spans over two and a half thousand years. Let us assume that the Universe began from something like the *Big Bang*. What happened before it began expanding? Sure, we see the fingerprint of Creation in many ways after this point, but what about before it took place? The Torah explains that God existed in 'the void' before Creation as an *Intelligence* that understood 'the Light'. This *Presence* or *Force* existed in the beginning, with God, as God. God merely spoke[108] and His Word created the Universe. This Word, while with God in the beginning, made all things. The *Being* that initiated the *Big Bang* was there in the beginning, and He[109] will be there for the end. He is, "the Alpha and the Omega, the First and the Last."[110]

---

108 "And God said" (Genesis 1:3, 6, 9, 14, 20, 24).

109 Note that ancient Hebrew, as recorded in the Torah and Old Testament is sex agnostic, whereby "Man" infers both sexes men and women.

110 "He who was seated on the throne of Heaven said, 'I am the Alpha and the Omega, the

In the *Bible*, this intricate detail continues. This book tells us that all things were made through God.[111] Christianity culminates the Creator of the Universe's discussions with humanity. As the chosen remnant to the Jewish people, and then to all people who profess Christ, throughout history, it differentiates itself from other religions. The *Bible*'s history presents tremendous detail and description, looking at science, rulers, behavior, and people across thousands of years, with names of people, places, dates, and timelines. It is an archeological masterpiece. It is an owner's manual on the human condition, problems, and ways in which mankind should aspire; it is a prophetic glimpse into the future, supported by history and mapping end times. In the following sections, we explore the scientific, historical, and prophetic accuracy of the *Bible*.

A. Scientific Accuracy

Many people assert that Genesis chapters 1-2 demonstrate definitively that the *Bible* is contrary to modern scientific thought. For example, atheist Stephen Hawking claims, "Before we understand science, it is natural to believe that God created the Universe. But now, science offers a more convincing explanation."[112] Many people fall into this deception. However, even though the Jewish Torah dates back several thousand years, well before any popular notions of modern science, it holds profound scientific

---

Beginning and the End. To him who is thirsty I will give drink without cost from the spring of the water of life. He who overcomes will inherit all this, and I will be his God and he will be My son" (Revelation 21:6-7).

[111] "Through him all things were made; without him nothing was made that has been made" (John 1:3).

[112] Hawking, Stephen, speech at Caltech, Pasadena, CA, April 16, 2013. Reference transcript, http://www.theory.caltech.edu/people/preskill/, accessed March 15, 2017.

accuracy. As we read just the first two chapters of Genesis, we see tremendous detail in both the timing of events and the processes, or ways, in which they occur.

The more we discover over the course of time, the more scientific analysis supports the events found in Biblical scripture. Books like *The Evidence Bible*[113] or Harvard theoretical physics' *Amazing Truths: How Science and the Bible Agree*,[114] illustrate a plethora of examples, for further reading. As we learn more about our physical Universe, this new learning builds our faith and enhances further credence to the *Bible*. While it may take delving into the Hebrew to understand the language and various aspects of history for the most accurate context for critical understanding, at a high level, for a society likely without even a telescope, the author of Genesis portrays his findings remarkably accurately. While we could pull from numerous examples illustrating scientific accuracies throughout Scripture, I look at what some might be considered to be the most complex, Creation.

In verse one of the book of Genesis, we see formation of the Universe, including all the heavens and earth. These records predate scientific tools, with scientists now recognizing the underlying implications only with modern science and tools like the Hubble telescope and non-visible spectrum. The *Torah* describes the early earth as dark and empty, formless, and lifeless. This coincides with expectations scientists hold about looking at other planets and their molecular composition (and likely abundance of hydrogen and carbon on Earth).

---

113 Comfort, Ray, *Irrefutable evidence for the thinking mind,* The Evidence Bible, ReadHowYouWant, 2011.
114 Guillen, Michael. PhD, Amazing Truths: How Science and the Bible Agree, Zondervan, 2016.

The text shares the story of Creation in a series of "days," which in Hebrew, may refer to an open period of time, not necessarily twenty-four hour periods. [115] Thus, on the first "day," we see the advent of the Earth's rotation in relation to the sun, with day and night. On the second "day" or period, we see separation and condensation of water vapor, pulling moisture to the ground and pulling it up higher in the atmosphere. It was at this time that basic life might have begun, with organisms living off photosynthesis and producing oxygen, protected from the sun's rays by an atmospheric layer. [116] This corresponds to evidence for a theory characterizing major environmental and atmospheric change referred to as the *Great Oxygenation Event*, where the Earth moved from being largely carbon dioxide, methane, and nitrogen rich to a plethora of oxygen. This is consistent with geochemists' recent research looking at ancient rocks. [117] Next, we see tectonic activity with the rising of dry land, allowing the waters to settle, bringing about plant life and vegetation. The next period, the thick atmosphere begins to clear, moving from a translucent layer to a transparent one, showing forth the night sky and sun. [118] In the fifth period, we see maritime life and birds of the air, predating God's Creation

---

[115] The Hebrew word found in Genesis that has been translated as "day" is םוֹי or Yowm. It can be defined as a 24-hour period or an open period of time. Source: http://judaism.stackexchange.com/questions/10079/what-is-the-meaning-of-yowm-in-bereshit, December 12, 2013.

[116] Lane, Nick, First Breath: Earth's Billion Year Struggle for Oxygen, New Scientist, February 5, 2010.

[117] Canfield, Donald, Oxygen: A Four Billion Year History, Cloth, January 2014.

[118] The Hebrew word used in Genesis 1:16 for, "God made two great lights," is Asa, which according to Strong's concordance, means, "Accomplish," as an accomplishment completed in the past. The creation of the sun and the moon referenced in verse 16 goes back to verse 1, when God made the heavens and the earth.
Strong, James, Strong's Greek and Hebrew Dictionary of the Bible, Aug 2011.

of land animals subsequently followed by humankind, both male and female.

Table[119]

Day 1: Verses 1-5 **Hadean Eon** – Age when oceans formed and atmosphere became translucent.	Day 2: Verses 6-8 **Archaen Eon** – Age when water cycle and oxygenated atmosphere were established.
Day 3: Verses 9-13 **Proterozoic Eon** – Age when continents formed, combined with Paleozoic Era with plant life on land.	Day 4: Verses 14-19 **Paleozoic Era** – Age consisting of six periods when super-continental masses shifted and reassembled away from the poles, with massive tectonic transformation, and created several still-existing mountain ranges.
Day 5: Verses 20-23 **Mesozoic Era** – Age when life from the sea thrived and ultimately led to birds.	Day 6: Verses 24-31 **Cenozoic Era** – Age when modern mammals and humans developed.

Day 1 (Verses 1-5, below) represents the Hadean Eon, which is the first age of the Earth's formation, from aggregation of matter to when the oceans formed and the atmosphere became translucent.[120] This era is thought to have transitioned from continued formation of our solar system with gas and dust from supernovas around our galaxy[121] to extreme volcanic activity with molten crusts forming around the planet, and vast deposits of asteroids and meteorites.[122]

---

[119] Christian, Jeremy, *Genesis Creation Story is Scientifically Accurate*, accessed on Dec 12, 2013 at: http://headlyvonnoggin.hubpages.com/hub/Genesis-Accurately-Describes-Geological-and-Biological-Formation-of-Planet-from-Surface-Perspective

[120] Rafferty, John, *Hadean Eon*, Encyclopaedia Britannica, June 2013, accessed https://www.britannica.com/science/Hadean-Eon

[121] Bengtson, S., *Early Life on Earth*, Nobel Symposium 84. Columbia University Press, New York, 1994.

[122] Kisak, Paul, *Abiogenesis*, Natural Processes For The Origin of Life, CreateSpace, 2016.

In the beginning God created the heavens and the earth. Now the earth was formless and empty, darkness was over the surface of the deep, and the Spirit of God was hovering over the waters. And God said, "Let there be light," and there was light. God saw that the light was good, and he separated the light from the darkness. God called the light "day," and the darkness he called "night." And there was evening, and there was morning - the first day.[123]

Day 2 (verses 6-8) represents the <u>Archean Eon</u>, which is the age establishing the Earth's crust, the water cycle, and oxygenated atmosphere.[124] This era is thought to have formed photosynthetic bacterial colonies, which are now found as ancient fossilized stromatolites in South Africa and parts of Australia. They still grow and thrive in Australia today.[125]

And God said, "Let there be a vault between the waters to separate water from water." So God made the vault and separated the water under the vault from the water above it. And it was so. God called the vault "sky." And there was evening, and there was morning—the second day.

---

[123] The Hebrew word for "day" in all six periods of creation from Genesis chapter one is "Yom". The first four are used directly, the last two used preceded with an article.

[124] Rafferty, John, *Archean Eon*, <u>Encyclopaedia Britannica</u>, June 2016, accessed https://www.britannica.com/science/Archean-Eon

[125] *Stromatolites of Shark Bay: Nature fact sheets*, WA Department of Environment and Conservation. Government of Western Australia. Retrieved September 3, 2011.

Day 3 (verses 9-13) represents the <u>Proterozoic Eon</u>, which is the age when many geologists feel the continents formed, many of our earliest mountains rose, and plant life on land began to take root. Early fossils identify this era.[126] This is also consistent with what one might interpret with the Biblical account.

And God said, "Let the water under the sky be gathered to one place, and let dry ground appear." And it was so. God called the dry ground "land," and the gathered waters he called "seas." And God saw that it was good. Then God said, "Let the land produce vegetation: seed-bearing plants and trees on the land that bear fruit with seed in it, according to their various kinds." And it was so. The land produced vegetation: plants bearing seed according to their kinds and trees bearing fruit with seed in it according to their kinds. And God saw that it was good. And there was evening, and there was morning - the third day.

Day 4 (verses 14-19) represents the <u>Paleozoic Era</u> when the continents moved from being a central mass to balancing between the poles. Here, early simple marine life flourished with the early invertebrate (ie: arthropods) with many scientists surmising also transition to early vertebrate land animals.[127]

---

[126] Windley, Brian, *Proterozoic Eon*, <u>Encyclopaedia Britannica</u>, June 2013, accessed https://www.britannica.com/science/Proterozoic-Eon

[127] Robison, Richard, *Paleozoic Era*, <u>Encyclopaedia Britannica</u>, September 2015, accessed https://www.britannica.com/science/Paleozoic-Era.

And God said, "Let there be lights in the vault of the sky to separate the day from the night, and let them serve as signs to mark sacred times, and days and years, and let them be lights in the vault of the sky to give light on the earth." And it was so. God made two great lights—the greater light to govern the day and the lesser light to govern the night. He also made the stars. God set them in the vault of the sky to give light on the earth, to govern the day and the night, and to separate light from darkness. And God saw that it was good. And there was evening, and there was morning—the fourth day.

Day 5 (verses 20-23) represents the <u>Mesozoic Era</u>, showcasing when life from the sea expanded and thrived, leading to birds of the air. This includes further development of plant groups and the Jurassic Period of dinosaurs.[128]

And God said, "Let the water teem with living creatures, and let birds fly above the earth across the vault of the sky." So God created the great creatures of the sea and every living thing with which the water teems and that moves about in it, according to their kinds, and every winged bird according to its kind. And God saw that it was good. God blessed them and said, "Be fruitful and increase in number and fill the water in the seas, and let the birds increase on the earth." And there was evening,

---

[128] Tang, Carol, *Mesozoic Era*, <u>Encyclopaedia Britannica</u>, June 2013, accessed https://www.britannica.com/science/Mesozoic-Era

and there was morning - the fifth day.

Day 6 (verses 24-31) represents the <u>Cenozoic Era</u>, with development of modern mammals and humans. It brings us to the present with existing continental formations, largest mountains, and despite numerous extinctions, the flora, fauna, and animal life that we see today.[129]

> *And God said, "Let the land produce living creatures according*
> *to their kinds: the livestock, the creatures that move along the*
> *ground, and the wild animals, each according to its kind." And it*
> *was so. God made the wild animals according to their kinds, the*
> *livestock according to their kinds, and all the creatures that move*
> *along the ground according to their kinds. And God saw that it*
> *was good. Then God said, "Let us make mankind in our im-*
> *age, in our likeness, so that they may rule over the fish in the sea*
> *and the birds in the sky, over the livestock and all the wild ani-*
> *mals, and over all the creatures that move along the ground." So*
> *God created mankind in his own image, in the image of God he*
> *created them; male and female he created them. God blessed them*
> *and said to them, "Be fruitful and increase in number; fill the*
> *earth and subdue it. Rule over the fish in the sea and the birds in*
> *the sky and over every living creature that moves on the ground."*
> *Then God said, "I give you every seed-bearing plant on the face of*
> *the whole earth and every tree that has fruit with seed in it. They*

---

[129] Berggren, William, *Cenozoic Era*, <u>Encyclopaedia Britannica</u>, October 2015, accessed https://www.britannica.com/science/Cenozoic-Era

will be yours for food. And to all the beasts of the earth and all
the birds in the sky and all the creatures that move along the
ground—everything that has the breath of life in it - I give every
green plant for food." And it was so. God saw all that he had
made, and it was very good. And there was evening, and there
was morning - the sixth day.

This account of Creation from the *Bible* and *Torah* is just one small example of hundreds of details found within the book. A critical eye shows credibility in the contextual understanding of the literal writings shared within it. Indeed, this example just touches upon the first verses of the first chapter, of the first book. I encourage everyone to read them, to ascertain additional scientific truths. Such detail in the texts affirm the sources and validate them as trustworthy for understanding scientific truths around us.

B. Historical Accuracy

The *Tanakh* and *Bible* demonstrate historical accuracy. They compile a multitude of stories, each filled with details comprising names of people and rulers, locations of cities and events, and descriptions of each. Thousands of discoveries have been made validating references throughout Scripture. While I list just a few of note, volumes of works can be found detailing more.[130] This validation of specificity sets the *Bible* apart as one of tremendous historical significance and relevance.

---

[130] There are several books and essays on this topic, so I do not delve into it here. For a more complete look at this topic, try McDowell, Josh, <u>Evidence That Demands A Verdict, Life-Changing Truth for a Skeptical World</u>, Thomas Nelson, 2017.

Pool of Siloam (discovered in 2004) – This is the pool where the apostle John (John 9:1-11) shares that Jesus healed the blind man. While the location of this site was clearly misunderstood until its discovery, workers seeking to fix a water pipe found it beneath part of Jerusalem. Two millennia of municipal growth hid its location, being consistent with the Biblical record, describing it as a means to tunnel spring water into the city, dating back to the time of Hezekiah around 730 B.C.[131]

Dead Sea Scrolls (discovered 1946-1958) – over 972 scrolls written from approximately 250 B.C. to 70 A.D. in several caves found about a mile north of the Dead Sea and include manuscripts of the Biblical canon (40%) as well as other Jewish-based writing (30%), and manuscripts from a Jewish sect. Most of the documents consist of parchment, but some are papyrus and bronze, written in Hebrew, Aramaic, Greek, and Nabataean (an early Aramaic language). They provide tremendous historical evidence of the consistency of the writings. The Appendix also lists several ancient texts referencing Biblical events.

Tel Dan Stela (discovered 1993-1994) – this inscribed stone dating back to around 950 to 750 B.C., tells a tale[132] in Aramaic of how an unnamed king defeats and kills the king of Israel, and the king from the House of David. It cross-references both the House of Israel and King David, and validates a timeline that coincides with the Biblical account of the Davidic reign.[133]

---

[131] "…Hezekiah's reign, all his achievements and how he made the pool and the tunnel by which he brought water into the city…" (2 Kings 20:20).

[132] A translation can be found at: http://en.wikipedia.org/wiki/Tel_Dan_Stele (Dec. 13, 2013).

[133] This original inscription mentions that Hadad made this ruler king, where Ben Hadad

Assyrian Palace (discovered 1928-1935) – The Assyrian King Sargon II (who replaced Shalmaneser V) and King Sennacherib (Sargon II's son and successor) recorded historical events back to about 720 B.C. on the palace's walls.[134] These descriptions validate battles for the fall of Samaria135 and the defeat of Ashdod.[136]

Hittite Kingdom (discovered 1834-1912) – After centuries of no archeological support, a wide range of discoveries in Boghazkoy, Turkey, consisting of buildings, sculptures, and over 2,500 cuneiform tablet pieces, validate the existence of the Hittite kingdom dating back to 2000 B.C., and share much about their culture and events.[137] This validates numerous references in the *Bible* on the impact and conflict associated with the Hittites.

Ebla Tablets (discovered 1974-1975) – over 17,000 clay cuneiform tablets were found near Ebla, Syria, dating back hundreds of years before Moses (around 2300 B.C.) validate an era of literacy, meaning Moses could have written the Law, and cross-reference Biblical terms, such as *Canaan* (a Semitic-speaking region in the Ancient Near East during the late 2nd millennium B.C.) and *tehom* (literally the Deep or Abyss, refers to the Great Deep of the primordial waters of Creation in the *Bible*). They are held in the

---

(son of Hadad) was king in Aram and turned his rule over to Ben-Hadad II, who ruled Aram around 900-895 B.C. This is consistent with the battles described in 2 Kings 13-14, where Ben-Hadad succeeded Hazael, after battles against the Kings of Israel and Judah, around 800 B.C.

[134] Poebel, Arno, *The Assyrian King-List from Khorsabad,* Journal of Near Eastern Studies, Volume 1, Number 3, pages 247-306. July 1942. See also: Loud, Gordon, *Khorsabad, Part 1: Excavations in the Palace and at a City Gate,* Oriental Institute [University of Chicago] Publications, 1936, Vol 38.

[135] 2 Kings 17:3-6, 24; 18:9-11

[136] "In the year that the supreme commander, sent by Sargon king of Assyria, came to Ashdod and attacked and captured…" (Isaiah 20:1).

[137] Kurt Bittel, Hattusha: The Capital of the Hittites. 1970. See also: E. Akurgal, The Art of the Hittites (English translation 1962).

Syrian museum of Aleppo today.

Tablet of Pontius Pilate (discovered in 1961) – Up until its discovery, Biblical antagonists denied the existence of the Roman governor of Judea, Pontius Pilate, the ruler who ordered Jesus' crucifixion. However, archeologists found a limestone block inscribed with his name and title near Maritima, Israel, dating back to 29-32 A.D.[138] It can now be found at the Israel Museum in Jerusalem.[139]

This represents just a small fraction of the archeological evidence[140] that repeatedly validates the *Bible* as a source of credible historical information unlike any other.

C. Prophetic Accuracy

The *Bible* also demonstrates its credibility as a source of wisdom and reproach through its prophetic accuracy. Although people frequently overlook them, the *Bible* lays claim to over one thousand prophecies. These glimpses of the future range from a short duration of a day to over two-thousand years. They include battles won or lost, rulers, deaths and means of death, famines, periods of prosperity, destruction of cities and buildings, rebuilding of temples, births, etc. These prophecies fill both the Old and New Testaments with such frequency, that similar to the scientific and historical records, I just pull a few references. However, what is unique about prophecy is the spiritual implications, whereby such accuracy further validates the Universe as an *open-system* and the divine nature accredited to

[138] http://www.Bible-history.com/empires/pilate.html (Dec 13, 2013).
[139] See also: http://www.imjnet.org.il/, reference 'Pontius Pilate'. 2013.
[140] See Appendices for a list of other references

the associated authors, or sources, of prophecy.

With the accuracy of the predictions, the *Bible* further differentiates itself from other religious and secular texts, affording it divine appeal and validation. According to one biblical scholar,[141] there were thirty-three prophecies fulfilled in just the last twenty-four hours of Jesus' life, with over three-hundred prophecies fulfilled by Jesus' birth, life, and death. Researchers calculate that the odds of fulfilling forty-eight specific, of these three-hundred, prophecies is 1 x 10^{157}.[142] For the skeptic, I realize that prophecies held and realized in the past and only self-referenced may not be too convincing, so I only list a few. The references, herein, predate all but a handful of the oldest books by over 800 years,[143] making it impossible to reference externally, so instead, the focus is on end times. There are around five-hundred various interpretations of prophecies conveyed in Scripture, which have yet to take place.

Date for the ministry and fulfillment of Christ – In Daniel 9:25-27, in about 490 B.C. (during the reign of Xerxes's son, Darius, in the first year of his rule over Babylon[144]) divine revelation is given to Daniel by the angel

---

[141] Mark Hitchcock, a Christian pastor, author, and leading speaker on the topic of Biblical prophecy. See also: http://marklhitchcock.com/about/, September 2013.

[142] Ankerberg, John, Weldon, John, and Kaiser, Walter, The Case for Jesus the Messiah, Harvest House, Eugene, Oregon, page 21, 1989.

[143] The Book of Kells, thought to have been created by Celtic monks around 800 A.D. contains the four Biblical gospels and is one of the oldest surviving books in the world, estimated at being just over 1200 years old, with the Iliad recently dated at around 750 B.C. "Linguistic evidence supports date for Homeric epics". BioEssays. 35: 417–20. 2013-02-18. doi:10.1002/bies.201200165. Retrieved 6/22/2016:
http://onlinelibrary.wiley.com/doi/10.1002/bies.201200165/full

[144] Daniel served Nebuchadnezzar following the capture of Babylon by Cyrus in 539 B.C. where Darius I (550-486 B.C.) took rule over Babylon about two years after Nebuchadnezzar III's Babylonian revolt in about 522 B.C. This occurred prior to Darius' defeat by the Greeks at the Battle of Marathon in about 490 B.C. Goold, GP, Harvard

Gabriel that the Christ will be 483 years after a decree to rebuild Jerusalem, which was made in about 457 B.C. This places Jesus' fulfillment of Scripture at about 27 A.D. It reads, "Know and understand this: From the issuing of the decree to restore and rebuild Jerusalem until the Anointed One, the ruler comes, there will be seven sevens, and sixty-two sevens. It will be rebuilt with streets and a trench, but in times of trouble. After the sixty-two sevens, the Anointed One will be cut off, and will have nothing… He will put an end to sacrifice and offering." Note, 62 sevens plus 7 sevens is 69 sevens, which with sevens being days, is 483 years (69*7).

Destruction of the temple in Jerusalem (70 AD) – Jesus boldly proclaims that the massive temple in Jerusalem would be torn down, stone by stone.[145] This occurred approximately forty years later, in 70 A.D. While frequently overlooked, no authors in the *Bible* record this historical fact. Interestingly noted, one of Jesus' disciples commented on the magnificence of the temple, right before Jesus laments on how the Jews will be killed and the temple knocked down, with no stone standing on another.[146] It was so striking that in three gospels, the authors each record Jesus' statement. Jewish tradition, politics, commerce, and social discourse centered in the temple. Rebuilding of the original temple completed in about 515 B.C., with the ruler during Jesus' time, Herod, just finalizing a massive refurbishment, which

---

Studies in Classical Philology. Harvard University Press. pp. 112–115. 1972. Retrieved Google books, December 20, 2013.

[145] "Do you see all these great buildings? Replied Jesus. "Not one stone here will be left on another; every one will be thrown down" (Mark 13:2, also referenced in Luke 19:44 and Matthew 24:2).

[146] "Look, Teacher! What massive stones! What magnificent buildings!" (Mark 13:1), coupled by, "The days will come upon you when your enemies will build an embankment against you and encircle you and hem you in on every side. They will dash you to the ground, you and the children within your walls. They will not leave one stone on another, because you did not recognize the time of God's coming to you" (Luke 19:43-44).

brought it close to its original glory from the times of its building under King Solomon. Massive cut stones laid the foundation, being as large as 37 feet long, 12 feet high, and 18 feet wide,[147] with the temple building standing almost 30 meters high, sitting atop the Kidron Valley, far below. Today, many of these same stones can still be seen in the Kidron Valley. The most likely explanation for the authors in the New Testament to not include something as impacting to their culture as the destruction of the temple, especially after multiple people record Jesus' prophecy on the topic, is that it had not yet occurred. This point also provides credence to the timing of the New Testament writings, further supporting the timing of the writings being within one lifetime of Jesus' life, death, and resurrection.

Jesus' Entry into Jerusalem on a Donkey – In 519 B.C., the prophet Zechariah writes that the king of Jerusalem and Zion will come riding a colt, the foal of a donkey.[148] Each of the gospels (Matthew, Mark, Luke, and John[149]) share that Jesus rode into Jerusalem riding the foal of a donkey, an animal of peace, on Palm Sunday.

Jesus' Death and Resurrection – While Jesus predicts His own death and resurrection shortly before the event,[150] it is told many times prior. This is shared in the gospels. In Daniel,[151] around 530 B.C., Daniel predicts the

---

147 Josephus, The Works of Josephus, edited by Whiston, William, Hendrickson Publishers, 1987, chapters 8-10.

148 Zechariah 9:9-10

149 Matthew 21:2-9, John: 12:14, Mark 11:1-8; Luke 19:28-35

150 Matthew shares Jesus telling them, "As you know, the Passover is two days away, and the Son of Man will be handed over to be crucified" (Matthew 26:2); see also, "We are going up to Jerusalem, and the Son of Man will be betrayed to the chief priests and the teachers of the law. They will condemn him to death and will turn him over to the Gentiles to be mocked and flogged and crucified. On the third day he will be raised to life!" (Matthew 20:18-19). See also Matthew 17:22-23 and Matthew 16:21.

151 Daniel 9:25-26

Anointed One will come and be killed.

Jesus' Role or Purpose - the prophet Zechariah (519 B.C.) states[152] that God will remove the sins of the people in one day through His servant, giving this servant the messianic title used in Jeremiah.[153] This is also presented in each of the gospels of the New Testament, approximately 550 years later.

While these and many other prophecies are shared throughout the *Bible*, with the associated events recorded elsewhere in history, too, what is possibly more compelling is how the *Bible* describes the times thought still to come. In an era without wireless communication, motorized vehicles, transponders, and electronic payment mechanisms, the depictions for the last days now seem all too surreal.

Galactic Transformation – Due to what sounds like a huge earthquake, the sun will turn black, the moon turned to red, and the stars will fall to the earth, and everyone from the wealthy to the poor will retreat into caves and hide behind rocks in the mountains.[154] A third of the earth will be burned up from hail, fire, and blood hurled down upon the earth. A huge flaming object resembling a mountain will be hurled from the sky into the sea, killing a third of sea life and destroying a third of all ships. Another star like a blazing torch will fall from the

---

[152] "Listen, O high priest Joshua and your associates seated before you, who are men symbolic of things to come: I am going to bring my servant, the Branch. See the stone I have set in front of Joshua! ...and I will remove the sin of this land in a single day" (Zechariah 3:8-9).

[153] "The days are coming," declares the Lord, "when I will raise up to David a righteous Branch, a King who will reign wisely and do what is just and right in the land. In his days Judah will be saved, and Israel will live in safety. This is the name by which he will be called: The Lord our Righteousness" (Jeremiah 23:5-6).

[154] Revelation 6:12-16

sky, making a third of the world's rivers deadly to those who drink from them.[155]

This may resemble several large meteorites, reference nuclear missiles, or possibly a combination of both. It seems descriptive, again, from the perspective of a first-century vision. While disastrous for humanity, it is not universally uncommon, with comets like ISON[156] popping up somewhat suddenly in 2012, and several other comets moving along similar trajectories toward Earth in the coming years.

Military Control - A plague of locusts swarm the Earth that look like horses prepared for battle with breastplates of iron and wings that thunder very loudly, with tails that can sting people who do not have the seal of God on their forehead that results in the pain of a scorpion.[157]

While time will tell, this could be any number of things, but currently sounds like an army of flying drones controlled by any one of a variety of national, global, or corporate powers. Each could include a form of electric or similar weapon to police the global masses.

Prophets in Jerusalem - Two prophets will prophesy in Jerusalem for about four years, with the power to control precipitation, causing a drought for the whole period. The ruler of the time will have them killed, and their bodies will be left in the streets of Jerusalem for three and one-half days to be mocked by the world. However, after three days, they will rise again until God takes them into heaven, which will coincide with a big, localized

---

[155] Revelation 8:7-8

[156] ISON was a highly touted comet about 5 kilometers in diameter shooting right in front of the earth before boomeranging around the sun (November 28, 2013) flying even closer to earth on its continued journey.

[157] Revelation 9:1-11

earthquake. The whole world will be watching in real time.[158]

While this prophecy seems reasonably clear, future generations will know its accuracy. With the trending political, economic, and spiritual climates, this may even happen in our lifetime… if people start posting about two prophets in Jerusalem that control rain, we may all want to drop to our knees before our Creator, who relayed such events 2,000 years in advance. As for people around the whole world, while when this was written, it must have seemed crazy, yet with the ubiquity of camera phones today, it is now likely that the whole world would watch such an event.

<u>Global Government</u> - Ten countries will come together led by seven rulers. It will be a one-world government, having authority over all people. One of the rulers will receive a fatal wound, but be healed. This will be widely publicized and will astonish the world. He will seize the opportunity to proclaim himself as deity and seek out the people of God to kill them. A second ruler will then appear, proclaiming that all people must worship the prior ruler who recovered from the fatal head wound. He will come as a prophet, likely claiming to be Jesus or a man of God, but he will deceive the world with many highly publicized miracles. He will mandate a global currency required by all people for trade, enabled by a mark on people's hand or forehead.[159]

We see creation of a global currency dating back to shortly after WWII, through the Hague, with all likelihood moving us toward a global government, depending on a variety of socioeconomic and geopolitical factors, possibly within the next generation or two.

---

[158] Revelation 11:3-13
[159] Revelation 13:1-17

114

Mark of the Beast - To buy and sell or trade, each person must receive a mark on his/her hand or forehead.[160]

This appears interesting and viable, with the proliferation of credit and debit cards, the Internet, and things like identity theft, coupled with satellite tracking, electronic banking, virtual currencies like Bitcoin or even PayPal, and smart-card technologies. In 1999, an independent British group identified that the best way to track people from satellite was by the back of their hands or the top of their foreheads. The world already has microprocessor chip cards embedded into dogs and prisoners for tracking,[161] so the small jump to complete this prophecy is to require it on all people and to expand it to include a unique ID and electronic purse. Increasing financial concerns, fear of criminal acts and needs for increased accountability, and fraud including identity theft, make this once far-fetched prophecy now seem all-too tangible.

While these examples may have once appeared unrealistic or outlandish, with each advancing day, some 2,000 years *after* they were written, they are now extremely plausible. Again, scholars find dozens, if not hundreds, of other examples of still-unrealized prophecies in the *Bible*, depending on their interpretation. With modern technological advancements, these prophecies look more realistic today, than when they were written.

What power offers accurate visions of the future? These examples presuppose sobering insights of moral and spiritual reality. Each holds eternal consequences for our acknowledgement and consideration.

---

[160] Revelation 13:15-17
[161] Brian Brady, *Prisoners to be chipped like dogs*, The Independent, January 13, 2008.

Prophecy is not subject to the limited scope of the scientific method. It requires different rules of ascertaining the Truth, but these rules are simple. If God divinely inspired the *Bible*, the prophecies must be valid. If the prophecies are valid, they speak to the accuracy of the *Bible* and the associated impact of its wisdom into each of our lives. Similarly, if the prophecies are validated over time, the source of each prophecy must be true. Again, if you hold doubts about God, doubt your doubt and look again.

*   *   *

The *Bible* (Tanakh and New Testament) is unique and affords considerable credibility as a source of Truth. We see it presents scientific, historical, and prophetic accuracy. For thousands of years, it survives largely intact and preserved. Perhaps, it survived for divine reason. Despite unparalleled status, few take the time to read or understand it. Like the owner's manual for life, the *Bible* sits on shelves around the world. Rather than waiting until something in our life breaks, each of us will benefit by gleaning its insights. While picking it up to use it as a Band-Aid (or plastic medical strip) when confronted by trials in life can work on a limited basis, it is just a glimpse of the relationship, peace, and purpose God wants to have with each one of us on an ongoing daily basis. However, the *Bible* still offers considerably more, as it also contains moral and spiritual insights, which can also be identified and tracked. Scientific experiments can be performed, but our limited dimensional perspectives must change.

Authors of the *Bible* also explain a divine power, with promises to grant wisdom and blessing to even those that simply read particular sections.[162] As one reads it, also termed the Sword of the Spirit,[163] it cuts through subjective understandings to bring objective Truth. It claims we will grow in faith,[164] wisdom, and discernment by flipping through its pages.[165] Most importantly, however, it speaks to the deity of Jesus Christ and profound statements that Jesus is the *Force* that created the Universe. Read or listen to it in your native tongue.[166] God can handle our inquisition. In the next chapter, I explain that Jesus Christ and the Jewish, Muslim, and Christian scriptures, which teach about Him, reveal the nature of this God and are a trustworthy source of understanding for moral and spiritual reality.

---

[162] "Blessed is the one who reads aloud the words of this prophecy, and blessed are those who hear it and take to heart what is written in it, because the time is near" (Revelation 1:3).

[163] Ephesians 6:17

[164] "Consequently, faith comes from hearing the message, and the message is heard through the word of Christ" (Romans 10:17).

[165] "...namely, Christ, in whom are hidden all the treasures of wisdom and knowledge" (Colossians 2:3). "For in Christ all the fullness of the Deity lives in bodily form" (Colossians 2:9).

[166] For access to the Bible in a growing list of world languages, please reference Hosanna Ministries, http://www.faithcomesbyhearing.com, which, at the time of writing, offers the audio Bible in over 774 languages.

Part IV

Jesus

Jesus

The *force* of universal Creation offers tangible proofs for a divine Creator. Additionally, as we just saw, the *Bible* presents scientific, historical, and prophetic truth also illustrating that our Universe is an *open-system*, being influenced by external spiritual and physical forces. This purports credibility of supernatural origin. However, this collection of documents also presents tremendously more. Specifically, it identifies and explains this Creator. While the *Bible* uses many names for our Creator, they all lead us to the person of Jesus. In this Part, we delve into the identity of God and Jesus Christ and explain the resulting implications for each of our lives.

Interestingly, Christianity, Islam, and Judaism share the same historical definition or identity of God.[167] All three claim the same universal Creator, with shared ancestry up until the life of Jesus Christ. Jesus differentiates each. Jews do not believe Jesus was the prophesied Messiah or even a prophet. They hold Him as a mere mortal man, still waiting for the true messiah to come. Meanwhile, the Muslim <u>Qur'an</u> describes Jesus as the Messiah,[168] but interprets this role as inferior to God, agreeing with His life and teachings,

---

[167] "God witnesses that there is no deity except Him, and [so do] the angels and those of knowledge [that He is] maintaining [creation] in justice. There is no deity except Him, the Exalted in Might, the Wise." The Qur'an, Ali Imran, 3:18.

[168] "The Messiah, son of Mary, was not but a messenger; [other] messengers have passed on before Him. And his mother was a supporter of the truth. They both used to eat food. Look how we make clear to them the signs; then look how they are deluded." Qur'an: 5:75.

yet seeing Him simply as another chosen prophet, akin to Elijah and Mohammad. Only Christianity equates Jesus with God. The way these faiths view Jesus is mutually exclusive, meaning at most, only one view can be correct, and our very eternal existence may depend on our view.

Christian theology typically speaks of God in three persons, the *Trinity*, with the Father, the Son, and the Holy Spirit. God is the Father, with the Son being the *means* of universal formation, holding all things[169] in balance. It is the Son, Jesus, who took on a body and lived in the Eastern Mediterranean about 2,000 years ago, was crucified on the cross, and raised from the dead, and seen by over five-hundred witnesses.[170] The Holy Spirit is God's presence on this Earth in both the physical and spiritual realms after Jesus' death,[171] on the day of Pentecost.[172] Each of these represents unique personalities within the single God, which term is both a generic term for the *Being* who created and exists outside of our space-time-energy continuum, as well as a term for the Divine nature shared by the three persons of that *Being*.

---

[169] Indeed, the disciple John defines Jesus as the Word that God spoke that brought creation into existence, characterizing Jesus as, "In the beginning was the Word, and the word was with God, and the Word was God. He was in the beginning with God. All things were made through Him, and without Him nothing was made that has been made. In Him was life, and the Life was the light of men. The Light shines in the darkness, and the darkness has not overcome it... He was in the world, and the world was made through Him, yet 'the world did not know Him'... And the Word became flesh and dwelt among us, and we have seen His glory, glory as the only Son from the Father, full of grace and Truth" (John 1:1-14; ESV).

[170] "After that, he appeared to more than five hundred of the brothers and sisters at the same time, most of whom are still living, though some have fallen asleep" (1 Corinthians 15:6). "...which we have seen with our eyes, which we have looked at and our hands have touched... we have seen it and testify to it, and we proclaim to you the eternal life, which was with the Father and has appeared to us" (1 John 1:1-2).

[171] Jesus states, "I am going to send you what my Father has promised; but stay in the city until you have been clothed with power from on high" (Luke 24:49).

[172] Acts 2, 1-41; Pentecost is the 50th day after the Sabbath of the Passover and is also called the Feast of Weeks (Deuteronomy 16:10) and the Feast of Harvest (Exodus 23:16).

While a multitude of topics carry a study of God any number of directions, this Part simply lays a foundation that "The Word," which is the Second Person of the Trinity, "became flesh and dwelt among us."[173] This may seem like a fairy tale, but this is how the *Bible* depicts Him, with the physically limited Man simultaneously being the universal Creator.[174] Our lives were intended to rejoice in a divine dance, or *perichoresis,*[175] with Creation, resting in authority bestowed upon us by Jesus, that we might live in harmony, conviction, love, peace, and purpose that enriches others in community.

A. The Universal Creator Made Flesh

As shared earlier, God is "the one, sole, supreme Being, the Creator and Ruler of the Universe," who is, "eternal, spiritual, and transcendent, Creator and Ruler of all and is infinite in all attributes." [176] If one gives credence to the Spiritual reality presented throughout the *Bible*, as evidence suggests we should, Jesus' role and identity are essential to all humanity. His identity requires our diligent and informed understanding. As evidenced in history and shared throughout the *Bible*, He is of God and equal to God. This should transform our perspective every day of our lives.

---

[173] "The Word became flesh and made his dwelling among us" (John 1:14).

[174] "I form the light and create darkness, I bring prosperity and create disaster. I, the Lord, do all these things. I am the Lord, and there is no other" (Isaiah 45:6).

[175] A Greek word for the intimate relationship between the three roles or persons of God (Father, Son, Holy Spirit). Torrance, TF, The Christian Doctrine of God: One being Three Persons, T&T Clark, London, 1996. Kaple, Rob, Sermon at Grace Midtown Church, Atlanta, GA, USA, 2017.

[176] *God*, Dictionary.com, Unabridged. Random House, Inc.
http://www.dictionary.com/browse/god (accessed: August 24, 2017).

I do not delve directly into proofs of Jesus' deity here, but rather explain His identity in the context of Scripture.

As shared, Jews view Him as a man, Muslims views Him as a prophet,[177] and Sikhs acknowledge truth in His teachings. Only Christianity attributes Him with status as man and Creator, equality to God. He is very different from some transcendental state or all-encompassing force. The disciple John, who lived and traveled with Jesus for about two years, describes Jesus as the personification of Creation.[178] The *Bible* describes Him as multi-dimensional and existing outside time and space. The very nature of Jesus is the personified *Force* who created the Universe[179] with all its galaxies, solar systems, and planets simply out of His being. He simply *Is*.[180] All this brilliant creation was done through a *force* that is God's Word.[181]

The basic idea is that rather than man creating religion to reach God, and thereby being man-made, God creates a personal relationship with us through a tangible, palpable *Being* we can understand. Through the life of Jesus of Nazareth, this reality came into our world; Jesus was flesh and

---

[177] "And [remember] when I inspired to the disciples, 'Believe in Me and in My messenger [Jesus].' They said, 'We have believed, so bear witness that indeed we are Muslims [in submission to God]." The Qur'an Project, Qur'an, al-Ma'idah, 5:111. "The Messiah, son of Mary, was not but a messenger; [other] messengers have passed on before him. And his mother was a supporter of Truth." The Qur'an Project, al-Ma'idah, 5:76.

[178] "In the beginning was the Word, and the Word was with God, and the Word was God. He was with god in the beginning. Through him all things were made; without him nothing was made that has been made. In him was life, and that life was the light of men. The light shines in the darkness, but the darkness has not understood it" (John 1:1-5).

[179] "For in Christ all the fullness of the Deity lives in bodily form..." (Colossians 2:9).

[180] When Moses asked God His name, He replied, "I AM who I AM. This is what you are to say to the Israelites. 'I AM' has sent Me to you (Exodus 3:14). This title is also translated as 'Yahweh' or 'Jehovah'.

[181] "In the beginning was the Word, and the Word was with God, and the Word was God. He was with God in the beginning. Through Him all things were made; without Him nothing was made that has been made. In Him was life, and that life was the light of men. The light shines in the darkness, but the darkness has not understood it" (John 1:1-5).

blood and felt profound emotion, felt temptation, and lived like each of us. As Jesus tells His disciple, Thomas, after His resurrection, "Put your finger here; see my hands. Reach out your hand and put it into my side."[182] With Jesus' desire to meet Thomas in the midst of his doubt, He focuses on relationship, first with "His Father in heaven" as the source for His human strength, and then with humanity as the outpouring of His love. It shows God reaching out to humanity, not humanity trying to reach God. It is not about what we do, but what Jesus does for us. In fact, He knows the number of hairs on our heads and promises to take care of those who earnestly seek Him.[183] Jesus came to Earth in human form to overcome spiritual forces that bind us all and keep us from experiencing the fullness of God directly. Jesus was born, died, and was crucified, yet resurrected in physical form, showing humanity that there is not only spiritual, but physical resurrection. He walked among us, transforming each of the disciples and teaching us of God's love, and moral and spiritual realities.

When we engage in relationship with Jesus, a quality time of meditating in and on Him every day becomes critical. God designed us with a predisposition toward Him, and He is waiting for us to turn our hearts and minds toward Him, our Father in Heaven, for replenishment of energy (spiritual, moral, and physical) from the *Source* of all things, Jesus. At a previous employer, almost daily, I got so frustrated with my work situation, the petty politics and insecure actions of my superiors and co-workers, that

---

[182] Jesus states, "Put your finger here; see my hands. Reach out your hand and put it into my side. Stop doubting and believe" (John 20:27).

[183] To God, "Even the very hairs of our head are all numbered" (Luke 12:7). Jesus tells those at His sermon on the mount, "...do not worry about your life, what you will eat or drink; or about your body, what you will wear.

I frequently had to break out my *Bible* and pray before meetings to get 'refueled'. This got my focus back on Jesus and really worked in bringing me both peace and patience from an *Eternal Source*.

When we rely on our own strength, we drift away from God. However, His strength is immeasurably more than our strength.[184] Ideally, we would turn to God naturally, yet this may require our proactive focus.[185] When we rely on God's strength, we witness an amazing gift,[186] yet it is His intention for all humanity on an ongoing basis. However, as mentioned, this God wants each of us to be subject to Him. "It is not those who call themselves Jew, Christian, or Muslim, that will inherit Paradise, but those that follow the Word of God."[187] His grace is made perfect in our weakness; it is in our weakness that His glory is most revealed.[188]

God goes to great lengths to reach each one of us. Through both the mundane and the miraculous, each one of us holds our own unique story, which becomes our personal testimony. We then enter into a great family of believers from around the world and throughout time. God knocks on the door of my life, and on the door of your life.[189] Jesus proclaims to each

---

[184] "Therefore do not lose heart. Through outwardly we are wasting away, yet inwardly we are being renewed day by day. For our light and momentary troubles are achieving for us an eternal glory that far outweighs them all. So we fix our eyes not on what is seen, but what is unseen. For what is seen is temporary, but what is unseen is eternal" (2 Corinthians 4:16-18).

[185] "Do not be misled: 'Bad company corrupts good character'" (1 Corinthians 15:33).

[186] "So I say, live by the Spirit, and you will not gratify the desires of the sinful nature" (Galatians 5:16).

[187] The Qur'an, Al Baqarah, 2:113 (135, 136).

[188] "My grace is sufficient for you, for My power is made perfect in weakness" (2 Corinthians 12:9).

[189] "All who sin apart from the law will also perish apart from the law, and all who sin under the law will be judged by the law. For it is not those who hear the law who are righteous in God's sight, but it is those who obey the law who will be declared

of us, "Ask and it will be given to you, seek and you will find; knock and the door will be opened to you. For everyone who asks receives; he who seeks finds, and to him who knocks, the door will be opened." [190] If we pray to our Father in Heaven, He hears our prayers.

B. The Deity of Christ and the Disciples

There are many *gods*, but only one God. These *gods* are anything or anyone we value above our identity, or worth, in Christ. Examples include Buddha,[191] Mohammad, Krishna, a demon (whether claiming to be an angel of light or not) or other entity in the spiritual realm, fame, money, sex, drugs, power, possessions, etc. It is only when we give these people or things undue prominence and value that we make them bad, allowing them to divert our gaze from Christ and from our relationship with the living God. Jesus claimed to be God.

> *I am the way and the truth and the life. No one comes to the Father except through Me. If you really knew Me, you would know my Father as well. From now on you do know Him and have seen Him. (John 14:7)*

righteous. (Indeed, when Gentiles, who do not have the law, do by nature things required by the law, they are a law for themselves, even though they do not have the law. They show that the requirements of the law are written on their hearts, their consciences also bearing witness, and their thoughts sometimes accusing them and at other times even defending them.) This will take place on the day when God judges people's secrets through Jesus Christ, as my gospel declares," (Romans 2:12-16).

[190] "Ask and it will be given to you; seek and you will find; knock and the door will be opened to you. For everyone who asks, receives; he who seeks, finds; and to him who knocks, the door will be opened" (Matthew 7:7-8).

[191] Buddha never claimed to be anything more than a questing man. While I understand how his philosophy could endure, why followers ascribe him such deity is only a puzzle.

Whoever believes in the Son has eternal life, but whoever rejects the Son will not see life, for God's wrath remains on him. (John 3:36)

While not politically correct, it is not Christians who defined the rule that Jesus is the only way to God. God did through Scripture, the Jewish prophets, and through Jesus.[192] Whether we acknowledge Him or not, we must still give an account of how we lived our life while here on Earth. According to the *Bible*, Jesus is the only entity by which we can be saved.

In looking at the Biblical account of Jesus' disciples, history records a profound behavioral change. There is a miraculous transformation from a thrice denying Peter standing from a distance and the strong, charismatic Peter of the early church dying for his proclamation of a risen Jesus. Due to their actions, the lives of each of Jesus' disciples provide credence to Jesus rising from the dead. What happened?

From Scripture, it seems clear that each of the disciples thought Jesus was the prophesied Jewish Christ. However, their initial actions illustrate their understanding of common Jewish tradition that He was to overthrow the government and become a worldly king, like King David or King Solomon, and restore the Jewish people to prominence. They anticipated being members of His inner circle, with power and titles. Like people today thinking about what they will do when they win the lottery, they sat around debating what roles and stature they would have in this new kingdom.[193]

---

[192] "I am the first and I am the last; apart from me there is no God" (Isaiah 44:6).

[193] "A dispute arose among them as to which of them was considered to be greatest"

When Jesus entered Jerusalem for the last time, He was given the reception of the Jewish people's future leader. Like a war hero coming back from battle, thousands of people lined the road, waving palm branches and crying out praises.[194] The disciples thought prominence was almost upon them. Despite Jesus' blunt statements of warning, they had no idea that within the week, He was going to be beaten and crucified as a common criminal.

As Jesus spoke with them for the last time, depicted in the Last Supper, clearly they still had no idea what was about to transpire. Seeing their lack of realization that His life was about to end and that they were the chosen recipients of God's mandate, message, and new covenant, we see Jesus' human nature fully revealed in His sweating of blood.[195] Clearly, this shows His tremendous anguish and sorrow.[196] Still, He relies on His Father for strength to pick up the burden of humanity's sins and leave God's redemptive message to such a band of ignorant followers.[197] Furthermore, despite being told repeatedly His hour was at hand,[198] it must have been a tremendous shock to each of them when circumstances turned so abruptly. Peter, who pledges eternal loyalty to Jesus, also revealing his knowledge that Jesus is the long-prophesied Messianic Jew of the *Tanakh*, fearfully

---

(Luke 22:24).

[194] "A very large crowd spread their cloaks on the road, while others cut branches from the trees and spread them on the road..." [shouting] "Hosanna to the Son of David! Blessed is he who comes in the name of the Lord!" (Matthew 21:8-9).

[195] "And being in anguish, he prayed more earnestly, and his sweat was like drops of blood falling to the ground" (Luke 22:44).

[196] "...he began to be sorrowful and troubled. Then he said to them, 'My soul is overwhelmed with sorrow to the point of death" (Matthew 26:37-38).

[197] John 17:1-24.

[198] "As you know, the Passover is two days away, and the Son of Man will be handed over to be crucified" (Matthew 26:2).

denies any association with Him three times over the course of just a few hours. The fact that the *Bible* records such weaknesses should be taken into account. Continually, throughout Scripture, we see with raw and vulnerable clarity, the imperfect nature of man and the divine nature of God.

Something profound happened to each of Jesus' twelve disciples. History records that all but one of them were killed violently[199] for their efforts to proclaim the astounding fact that Jesus rose from the dead and promised eternal life to those who believe in Him. Many could have escaped persecution and death if they simply recanted this claim, yet they held firm. What would make each of them give up their own lives in such a violent way, unless they really believed? What could make them believe, especially after seeing Him killed as a common murderer? The *Bible* narrative speaks of many physical appearances of Jesus after He died on the cross; He was resurrected as He foretold, solidifying the disciples understanding of His deity. The fact that His disciples believed so strongly that they openly proclaimed Christ's teachings and resurrection to bring about their own murders speaks strongly to the truth of their proclamation.

While history boasts a long list of martyrs across a long list of both political and religious causes, each found their fate, because of their staunch beliefs; they believed in what they were doing. If they did not believe it but simply wanted others to believe it, would the movement have proliferated globally, whereas all other such movements die out? While many zealots die for their beliefs, what others have transformed the world to such an extent? If Jesus' divinity and resurrection were merely a ruse, it seems highly

---

[199] See Appendix for list of the demise of each of Jesus' apostles.

probable that such an illusion and lies would not have survived in the tight community in which the disciples lived, and Christianity would have died off long ago.

Jesus claimed to be the Messiah and God. Several proofs explain how to interpret such claims. While this logic is further developed in a variety of works,[200] I provide a brief overview. First, if Jesus said He was God, but knew He was not, He was a *deceiver*. For such lies, He must be evil, with resulting proofs stemming off this idea. Second, if Jesus was not God, but thought He was, He must have been a form of egomaniac and a *lunatic*. In either situation, others intimate with His life would have seen and known His state of mind, leaving little probability of any following or global transformation. Third, if Jesus did not claim to be God, but a manipulative early church or government made it up, then there is a *false historical narrative*. However, such an idea simply ignores history.

Briefly, first generation followers of Jesus were severely persecuted. It simply would not have been convenient for anyone to be a Christian until, at least Constantine, some three-hundred years later, yet despite people being severely persecuted and killed for their faith, it not only endured, but flourished. By the time the early church created the canon of the *Bible*, possibly around the time of Origen (300 A.D.), the religion was already widespread, despite continued social and bodily harm to followers. Numerous accounts of God from around the Mediterranean world fueled its growth, despite repeated efforts to crush it. As shared by the Pharisee, Gamaliel, if man-made, it would die after a generation.[201] Moreover, the

---

[200] McDowell, Josh & Sean, <u>More Than A Carpenter</u>, Tyndale Momentum, 2009.
[201] "Some time ago Theudas appeared, claiming to be somebody, and about four hundred

evidence of Christ's disciples and their commitment and teachings strongly opposes the idea that they would manipulate such a claim.

Finally, if Jesus thought He was God and is God, then the *Bible* and history are correct. *Jesus was and is God.* With this truth, let us all deepen our relationship with Him.

•

C. Professor Analogy

On the first day of class, a professor at a university tells his class that their entire grade is based upon three papers balanced out through the semester. He says that any late papers will receive an 'F'. The students understand this, but when the first midterm paper becomes due, invariably, there is a student that cannot get the paper in on time. When the student approaches the professor and explains the situation, the teacher concedes and offers a short extension. Word of the extension gets around the class.

By the time the second paper comes due, a handful of students fail to turn it in on time. The professor hears all the stories, "my dog ate it," "my kid brother flushed it down the toilet," "my uncle died," and so on, yet he grants each one of them a short extension. For the final exam, only a fraction of the class turns the paper in on time. While each person with a late paper expects an extension, none is given. Each of the students gets an *F* averaged into their semester grade. When this realization hits the

---

men rallied to him. He was killed, all his followers were dispersed, and it all came to nothing. After him, Judas the Galilean appeared in the days of the census and led a band of people in revolt. He too was killed, and all his followers were scattered. Therefore, in the present case I advise you: Leave these men alone! Let them go! For if their purpose or activity is of human origin, it will fail. But if it is from God, you will not be able to stop these men; you will only find yourselves fighting against God." (Acts 5:36-39).

students, one declares, "Professor, this is unfair! I demand justice!" To this the professor replies, "Actually, it is 'justice' that you are getting. What you want is 'grace,' but grace has already been given."

This is our reality today. Spiritual laws govern moral truths that impact each of us in our physical lives. Many people today accuse God of being unfair, saying that it is unjust for Him to punish us with death for our sins. Actually, it is just, and people want forgiveness and grace, not justice. In this analogy, as long as we are alive in this world, we are in the middle of the semester. We have time to choose, and, as I will explain in the next section, God gives us grace and an opportunity to choose forgiveness. However, once we die, it will be too late to choose, and God will judge those who have not accepted the gift of Christ.

D. Sacrifice for Sin

It may contradict human understanding to believe that the relational, loving *Force* that created our Universe would subject Himself to embodiment in a limited dimensional form, but this is exactly what the *Bible* teaches. Christ came into the dimension of Earth and lived a perfect life, so that we might have a sacrifice for our sins. Why do we need a spiritual sacrifice? This is a good question and one that speaks to our spiritual reality.

Scripture teaches that we need a sacrifice, because there is a spiritual law stating that unless we live a perfect moral life, we must die both a physical and a spiritual death. It looks to be a truth that exists outside our Universe's *open-system*. However, if we have a sacrifice that can fill this tremendous gap between our imperfections and God's perfection, we may have a spiritual life in heaven after our physical deaths. Thus, we see an old Jewish tradition spanning thousands of years, not coincidentally up until the time of Christ, where the Levites, the priestly tribe of Jews, sacrificed[202] countless animals for their blood. This did not bring permanent forgiveness, only temporary atonement. Thus, the killing of animals continued as an ongoing process, year after year. God did this to save humanity.[203]

With a singular event, the crucifixion of Jesus carries forgiveness to all humanity, throughout time, who choose to partake in it. In His perfection, God created a path, or process, for us to overcome the gap that all the animal sacrifices could not provide. In the sacrifice of God, Himself, as Jesus, He meets this spiritual law for those who believe in and accept Him.[204] That is why there is so much debate about Him being man and God.[205]

E. Smoking Gun

I recently watched a television show about a man captured by the Vikings. In one scene, he gains courage in the face of the Vikings' many questions, thinking their interest is based on a humanistic quest for

---

[202] Ironically, the Jews, who do not accept Jesus' divinity, stopped offering sacrifices shortly after Jesus' death, around A.D. 70, when the temple in Jerusalem fell. Per their beliefs, the Jewish people should still be sacrificing animals regularly, which is an interesting spiritual moral side note.

[203] "For God did not send his Son into the world to condemn the world, but to save the world through him" (John 3:17).

[204] "The reality, however, is found in Christ" (Colossians 2:17).

[205] "…that we may live peaceful and quiet lives in all Godliness and Holiness. This is good, and pleases God, our Savior, who wants all men to be saved and to come to a knowledge of the Truth" (1 Timothy 2:2-4).

knowledge and appreciation of his culture. However, in the subsequent scene, he learns that the Vikings simply manipulated him to reveal weaknesses in his homeland with secrets that would allow them to attack his people. As soon as his eyes open to their true motives, he tells them that everything from the previous scene was a lie. In recanting his prior statements, however, the Viking leaders realize the truth of his original stories. Thus, in the man's attempted deception, the authentic truth crystallizes.

So it is with true faith. Walking through the streets of Boulder and many other cities around the globe, I hear people talk about worshipping Buddha, dedicating months to the study of self-meditation, focusing on their own needs as the basis for attempting to find happiness and peace, and affirming every faith or religion outside of Christianity. Even smoking marijuana as a means of worshipping *Jah* is considered completely acceptable, if not trendy. However, as soon as one mentions the name *"Jesus,"* people automatically become agitated and presume the speaker is closed-minded and a hater of all things different. This depiction of spiritual oppression exists in many progressive and modern cities, albeit to a lesser extent in other parts of the world, which still seem more healthy and open to discussions of faith. Still, it is a wonderful spiritual proof, clarifying that Jesus is truly different. It is sweet to those that know it and bitter to those that do not.

As with the Vikings, this vehemence and hostility toward Jesus clarifies the Truth. The *Bible* teaches us about loving God, our universal Creator, and loving people, our neighbors. It explains that the person doing the most, even in the name of God and with God on the tip of their tongue; is an empty and a resounding gong, unless he or she loves abundantly.[206] It poses the question that Jesus is very different. While Christians sometimes hide behind subjective interpretations and limited understandings, this is true of all people and worldviews and should not be seen as a strike against just Christians. Renowned scientist and vocal atheist, Stephen Hawking, battled much of his life to disprove God. He pursued it with a religious fervor and zeal unfitting for a purely scientific pursuit. However, if he was truly an atheist, and he really believed God did not exist, with this life simply being without purpose or reason other than selfish interest, then why should he even care? Philosophically, a true atheist sees no ultimate value or purpose in life, with birth as meaningless as death, so effort directed against disproving (or proving) anything is just academic vanity.

As an atheist, the basic tenet sets that life holds no higher purpose. Hawking, and everyone else in humanity, lives and dies for absolutely no reason other than that we came into existence. Again, in atheism, each person creates purpose directly out of the depths of his or her heart, and every purpose is completely subjective to each person. Thus, an atheist's irrational lust to disprove the true Creator God supports spiritual reality and the existence and Truth of God. When one finds a smoking barrel, it is safe to assume that the gun just fired, meaning that because there is so much conflict around Jesus' love and identity, it shows the spiritual warfare taking place and truth associated with Jesus.

---

[206] "If I speak in the tongues of men and of angels, but have not love, I am only a resounding gong or a clanging symbol" (1 Corinthians 13:1).

F. Gravity

> *But now a righteousness from God apart from law, has been made*
> *known, to which the Law and the Prophets testify. This*
> *righteousness from God comes through faith in Jesus Christ to all*
> *who believe. There is no difference, for all have sinned and fall short*
> *of the glory of God, and are justified freely by His grace through the*
> *redemption that came by Jesus Christ. (Romans 3:23-24)*

Those who accept Christ and follow His teachings experience forgiveness of sin, but the effects still impact their lives. When actions that honor God become a conscious decision, as opposed to automatic and natural, it is a sign that our focus has fallen off Jesus. In such events, faith[207] and wisdom help keep our relationship with God from separating even further. If we do not restore this relationship, it may be detrimental to us not only spiritually, but also oftentimes physically. In these times, it is important for us to turn away from the course of events and actions that cause us to lose our focus and fall away. Oftentimes, pride keeps us from doing this, or we fall victim to the lie that God will not forgive or like us anymore. Yet, it is critical to restore relationship with our Creator, regardless of the situation. We cause the separation, yet He meets us in our place of need, enabling restoration. We must turn toward Him, or we may end up drifting even further away.

Imagine a ball. Let it be a soccer ball, a big red rubber ball, or whatever ball you want to create in your mind. It is suspended about ten feet above

---

[207] "…it is by faith you stand firm" (2 Corinthians 1:24).

the ground. What happens as its potential energy transforms into kinetic energy? It falls toward the ground. As well known, we call this force *gravity*. The only thing that is going to stop the ball from hitting the ground is an opposing force, taking it to a subsequent state of rest. This could be a table, a chair, or another obstacle. There are many forces in the world that could knock the ball off its perch, such as a breeze or someone bumping it. Eventually, though, without additional forces proactively raising the ball, it will hit a passive state of rest.

Now, this ball represents your spiritual life. Gravity is the force that pulls on us, as we go about our daily lives in an imperfect world. Each time we break a spiritual law, we fall a little bit; each moment in our day where we focus on our own abilities apart from God, *our ball* falls a little bit. The obstacles are many. Perhaps family, a spouse, friends, a job, looks, charm, money, power, social media (ie: Snapchat), or busyness blur our self-perception and make us unaware of our fall, yet it still occurs. Due to inherent imperfections in ourselves and in the world around us, in the absence of Christ, it is natural for us to eventually hit the *ground*, our spiritual ball finding rest. Gravity. Here, our spiritual self stays. The winds of life blow or some other force bounces our ball along the ground, rolling it into walls and rocks. We may even find a staircase that we bounce down, relinquishing more kinetic energy, bringing us to an even lower state of rest. It is inevitable, and over time, we expend our potential energy. We eventually hit the lowest state, at the bottom of our spiritual well.

The only thing that raises us is relationship with our Creator. Like new energy lifting the ball, when we focus on Christ with an earnest and sincere heart,[208] we defy *gravity*. In putting an end to our selfish desires, which we see as our ball falls, we seek eternal rewards and a renewing of our minds through Christ. While we must be a Christ-follower to receive this blessing, by God's grace, it does not mean that all people that follow Christ are always Christ-focused. Gravity affects everyone. While many mature Christians may appear to defy gravity, in reality, gravity continually pulls on them. They simply learned how to remain Christ-focused. As a result, their ball rises, overcoming the world's force with its temptations and challenges. Newer Christians may find their ball bouncing from ground to sky. Non-Christians seem to always be surprised to see a Christian's ball on the ground. While it is unfortunate when this happens, it speaks to the Christian's lack of focus and spiritual maturity and is not only possible, but likely for Christians during certain points throughout their lives.

There are just two directions. We can look toward Christ or toward the world and our own power. Despite our vain desire to focus on other things, our vision falls in just one of these two directions. Both are opposite from one another, and it is impossible to focus on both at once. However, God wants us to be *in* the world, just not *of* the world, so when we look to God, we see the world with God's heart, as if through the lens of God. God moves between us and the world, so we only see the world through Christ:

---

[208] "Do not conform to the pattern of this world, but be transformed by the renewing of your mind. Then you will be able to test and approve what God's will is - his good, pleasing and perfect will" (Romans 12:2).

The direction we look determines the direction we walk, spiritually speaking. Thus, in order to grow closer to God, we must look to Christ at all times, not only keeping the ball elevated, but even defying gravity and moving it up to a higher state of kinetic energy. This is why Jesus spent so much of His short, precious ministry in prayer and told us to pray ceaselessly and to seek God at all times and in all things.[209] When we look away from God, our ball falls. Does it sound impossible to continually focus on Christ? In Christ, all things are possible. Additionally, He calls us to glorify Him in everything we do.[210] Whether our work raises our ball or lowers our ball, it depends on our focus, heart, and intentions.

G. Faith

A friend once emailed me, "I do have a question though... Are your Christian beliefs and faith automatic or is it a decision that you make every day?" To answer this question, one must establish a common definition for the term "faith". As such, I define faith as, "a belief in the unseen, characterized by action in support of this belief."

True faith is about a relationship with the great "I Am" and is the antithesis of religion. God divinely inspired the tenets outlined in the Bible. It is not a "man-made belief system," but one inspired by the Creator of the Universe before time began. True Christianity is all about God reaching out to humanity, not humanity trying to reach God. It is not about what we do, but what God does. We, as humans on one speck of dust (Earth) in the

---

[209] "So whether you eat or drink, or whatever you do, do it all for the glory of God" (1 Corinthians 10:31).
[210] "Whether by word or deed, do all things in the name of the Father" (Colossians 3:17).

Universe, are subject to the reality of this Creator. We grow in faith not by following a set of rules, but by drawing into relationship with Him. All good relationships take time to build and work best upon a foundation of quality time and interaction.

If gravity always pulls one down, how does one rise above this natural and ubiquitous force? Simply stated, it takes effort. While relationship with our Creator can seem completely natural, all relationships require action from both parties. Furthermore, conflicts and forces of spiritual nature pull us away, just like gravity, from our focus on Christ. With free will, we all play our part in managing our choices. This is why it is so critically important to keep our focus on Jesus on a continual basis.[211] Unless I am consistent in prayer and seeking His will, my first inclination is to satisfy my own needs. This is childish. In seeing and hearing God and His truths, we build our faith. One way to grow in faith is simply to read or hear the words of the *Bible*.[212] This is why so many people encourage followers of God to read the *Bible* daily and to reflect upon Christ continually.

For Christianity to have eternal value, it must fight to avoid becoming a *religion*. Even in Scripture, not all issues are black and white, and some topics convey a subjective or temporal interpretation. What separates one person from relationship with our Creator may cause another person to draw closer. For example, for a person that controls his or her drinking, having a glass of wine may be acceptable and can even draw him or her

---

[211] "Be joyful always; pray continually; give thanks in all circumstances" (1 Thessalonians 5:16-18). While this is the goal, it is typically not easy to achieve, only being possible through God's grace.

[212] "Consequently, faith comes from hearing the message, and the message is heard through the word of Christ" (Romans 10:17).

closer to God. However, for an active or recovering alcoholic, even one sip of alcohol may be a categorical sin, pushing him or her away from God. I have a good friend, Mike, recovering from alcoholism (seven years sober). Knowing this fact, while I may be okay grabbing a drink with a different friend, because drinking takes Mike away from God, my drinking in front of him also causes me to sin. In knowing about his alcoholism, my drinking in front of him acts as a form of temptation, which we can avoid.

Followers of this objective reality must be sensitive to these truths, fighting in love to bring our own clarity and understanding, as well as carrying it to those around us. However, this does not mean that all topics are subjective or that subjective truth leads us closer to God; there are still objective truths and realities, as the Spirit of God shares.

H. Free Will

People often wonder why God, our Creator, sends *good people* to hell? This question fails to understand the spiritual law that dictates any imperfection contaminates the whole. Just as a little yeast, or leaven, in new dough makes it rise, one failure in moral law carries us into eternal separation from our Creator. While God is all-loving, He is also just and righteous. While slow to anger, He does have wrath.[213] He is not who we want Him to be, but rather He is who He is. It is our responsibility to figure out His identity.

As related in the book of Genesis (accepted by Jews, Muslims, and

---

[213] "In furious anger and in great wrath the Lord uprooted them from their land and thrust them into another land, as it is now" (Deuteronomy 29:28).

Christians) with the Fall of Adam and Eve in the Garden of Eden, we live in a spiritual war-zone. In looking at my own life, it makes sense to me, as my many imperfections continually surprise me. While I would like to think of myself as being inherently good, I have lived long enough to know this is both naïve and untrue. Perhaps, I am good compared to other imperfect people, but such comparisons fall short of God's objective, eternal perspective. Therefore, we must search and reflect upon our thoughts, deeds, and motivations with sober clarity. Comparing ourselves to those around us, rather than His perfection, offers a false narrative of *good*. All good things come from God.[214] No one is truly good, except God alone.[215]

An analogy to explain our spiritual state equates God with light.[216] If God is light, what is darkness? Darkness is an absence of light. Applying this analogy reveals darkness is an absence of God. Spiritual darkness is separation from God; it is a dominion of personified evil.[217]

God created the propensity for this embodiment of evil to exist, bound in this darkness, and roaming ceaselessly throughout the earth.[218] Objective reality is spiritual reality. To overcome the spiritual forces of evil in the

---

[214] "For everything God created is good, and nothing is to be rejected if it is received with thanksgiving, because it is consecrated by the Word of God and prayer" (1 Timothy 4:4), in the context of eating food and marriage.

[215] Matthew 19:17, Mark 10:18, Luke 18:19

[216] "God is light; in him there is no darkness at all...But if we walk in the light, as he is in the light, we have fellowship with one another, and the blood of Jesus, his Son, purifies us from all sin" (1 John 1:5, 7). See also: "The light shines in the darkness, but the darkness has not understood it" (John 1:5).

[217] "For he has rescued us from the dominion of darkness and brought us into the kingdom of the Son He loves, in whom we have redemption, the forgiveness of sins" (Colossians 1:13-14).

[218] "Satan answered the Lord, 'From roaming through the earth and going back and forth in it'" (Job 1:7). See also Job 2:2.

heavenly realms,[219] which frequently seem to control this world,[220] we must trust God for our physical reality.[221] God receives our thanks when we praise and honor Jesus in everything we say and do.[222] No matter how bad our circumstances may appear, God can pierce the darkness. Indeed, God's omnipresence can be anywhere we seek Him. If we ask, we will receive.[223] As light shines into a dark room, wherever God goes, the darkness flees. It is a simple spiritual truth. Thus, we should not worry or fear burdens and trials placed upon us. It is not as much about the things we have done or where and how we find ourselves in life. Instead, it is about where we are going, our hearts and minds, and whether we place our eyes on the Creator or the created.

People who live lives away from God often feel isolated from Him, thinking that their situation is somehow special or circumstances too dark for God. This is not true, as God can appear in any circumstance, to those who earnestly seek Him.[224] People who believe in God often feel they are

---

[219] "For our struggle is not against flesh and blood, but against the rulers, against the authorities, against the powers of this dark world and against the spiritual forces of evil in the heavenly realms" (Ephesians 6:12).

[220] "Jesus said, 'My kingdom is not of this world. If it were, my servants would fight to prevent my arrest by the Jewish leaders. But now my kingdom is from another place.'" (John 18:36) and referring to the devil, Paul writes regarding those that do not yet know God, "In their case the god of this world has blinded the minds of unbelievers, to keep them from seeing the light of the gospel of the glory of Christ, who is the image of God" (2 Corinthians 4:4).

[221] "Therefore I [Jesus] tell you, do not worry about your life, what you will eat or drink; or about your body, what you will wear" (Matthew 6:25).

[222] "And whatever you do, whether in word or deed, do it all in the name of the Lord Jesus, giving thanks to God the Father through Him" (Colossians 3:17).

[223] "This is the confidence we have in approaching God: that if we ask anything according to his will, he hears us. And we know that he hears us – whatever we ask – we know that we have what we asked of him" (1 John 5:14-15).

[224] Take Paul again, who authored many of the books (letters) in the New Testament of the Bible, and who was so zealous for God and Judaism, that he sought to persecute the

unworthy of forgiveness or not deserving of God's light. Of course, none of us deserve this; how could we? If we deserved His grace, it would not be called *a gift*, but *payment* or *wages*. The only *wages* we receive are "the wages of sin," which is *death*.[225] However, God gives those who accept it in humility and self-realization, grace and forgiveness through Christ's sin sacrifice, with His death on the cross. God created humanity with the ability and propensity to choose how they behave. We have free will to turn toward Him or to turn away from Him. People always ask me, "Why did God not just make us good and holy?" He wants our praise coming from all aspects of our being (mind, heart, body), not an unthinking ritual or mechanical action.

I. Sin and Pain

Many people think of pain as a bad thing, ascribing a negative connotation. However, our nervous system's ability to flash pain is wonderful. Without it, we would unknowingly hurt our bodies and possibly suffer irreparable damage. We tend to associate our distaste for pain with the situations that cause us to feel pain. Without pain, our bodies would fall apart quickly due to misuse and abuse. Have you ever touched a hot pan or dish? The natural reaction is to pull away from it quickly and then, maybe, run your finger or hand under cold water. Our complex nervous system

---

followers of Jesus to the extremity of death. His passion was so strong that he even obtained special papers to extend his authority outside his original domain. While a murderer of Christians, God later used him as a powerful representative of good and for His glory.

[225] This verse continues, appropriately, "...but the gift of God is eternal life in Christ Jesus our Lord" (Romans 6:23).

shoots near instantaneous warning, enabling us to change our behavior to avoid extensive harm.

Like our body's ability to feel pain, our knowledge of those things that we do that take our focus off God is good. *Sin* can be defined as anything that takes our focus off God. While many people think of guilt as a negative thing, knowledge of the sin causing it is essential for correcting an inappropriate action. Since sin is what separates us from God, our natural tendency should be to flee from it. It identifies moral realities, with our collective moral programming built into each of us. Guilt provides knowledge of sin similar to our nervous system's sensations of pain; it offers a strong warning that we are harming ourselves. Without this pre-programmed knowledge of sin in our hearts, we might miss awareness that our actions separate us from relationship with our Creator.

Society teaches us that rather than deal with guilt or accept responsibility for our condition, we frequently flee from or obstruct even our own self-awareness of our action(s). While there are objective sins, the moral reality is that there are many more subjective sins. These can frequently be subjective in nature to each person and correlate with a circumstance or situation.

As we grow older, we frequently accept such feelings of separation and isolation, continually seeking anything but relationship with the Creator to fill our void and restore us to peace. Not only are we rarely schooled in how to deal with it, but society tends to discount it, confusing us, and even promoting the very actions that caused our separation from God in the first place. Even talking about it is taboo and cause for alarm in many social circles, which only validates the spiritual warfare surrounding the topic.

146

In order to keep our bearing and path, we must judge ourselves to understand our heart and whether our gaze focuses on God *or* on ourselves, the world, and the created. Psychologists and psychiatrists delve into many of these issues, but do not always give an appropriate, or God-glorifying, interpretation. Doctors may prescribe a chemical fix for our moral pain. Our friends may tell us we need more self-focus or that being more active may help us. As many of us know, drugs, exercise, or self-focusing rarely fix the issue. In fact, not addressing the underlying reasons for guilt may make matters worse. Escapism through drugs or alcohol, as well as self-focused gratification,[226] also rarely eliminates feelings of guilt and sin. Many times, we must embrace and acknowledge the pain to identify where and how our steps take us off God's path, so we may change course.

J. Dying to Sin

As a child, I did not mind pain, but was terrified of blood. One day, when I was four, I knocked a pickle jar off the shelf and as it shattered into the floor, a shard of the glass carved out my toenail. While jolted by the initial shock, the cut severed the nerves, and I felt little pain. Additionally, I was not yet trained to know what it meant or how to respond. I began walking around the house, looking for my mother, but she was unknowingly hosing off the back porch. After a few minutes, I finally noticed the trail of blood following me, saw my toe, and immediately began to cry.

---

[226] Jesus proclaims, "If anyone should come after Me, he must deny himself and pick up his cross daily. For whoever seeks to find his life shall lose it, and he who loses his life for Me, shall find it" (Luke 9:23-27).

We respond similarly when our actions take our focus off God. We get so accustomed to the pain and feelings of guilt, that we accept them. We do the same harmful actions so many times that we become accustomed to the pain and no longer acknowledge it. Similar to me not feeling my toe, we sever our spiritual nerves. In rationalizing our actions, we tell ourselves our responses may even be good or healthy, rather than damaging to our souls. We reinforce a negative response, then deal with our moral repulsion through frequently flawed means. As a result, we ignore sin's warnings. When this happens, we no longer feel the pain or separation sin should cause and thus "die" to sin associated with our actions.

Renowned existentialist philosopher, Friedrich Nietzsche, claims that when we do something enough that we no longer feel guilt, we claim victory over the moral reality of our actions.[227] Like a sharp piece of glass cutting through nerves, dying to the feelings of guilt allows us to do and say things that separate us from the personal Creator of the Universe without clear self-awareness.[228] Rather than agreeing with Nietzsche's claim of victory, however, we should see it as a horrible loss. This moral reality clouds our knowledge and realization of sin; thereby, hindering our moral programming and making it easy to slip refocusing on God. However, developing a healthy relationship with our Source and Creator requires our refocus on Him following missteps. When we die to innate feelings of guilt, only God can restore our hearts to a childlike state; this is a frequent miracle experienced by many people who seek relationship with Him.

---

[227] Nietzsche, Friedrich, <u>The Gay Science</u>, Vantage Books, New York, 1897.
[228] Interestingly, Nietzsche, an existentialist writer, claims in <u>The Gay Science</u>, that this as man's triumph.

We should all seek to preserve our innocence. Our culture, meanwhile, tells us to embrace actions that take our innocence from us. We must employ wisdom in the choices we make, the shows we watch, the materials we read, the people with whom we spend time, the things we do, and the things we say. I remember speaking with a seven-year-old girl one day about her desire to watch a violent television show. She replied that it was okay, because her mom lets her watch whatever she wants, and she sees people die all the time. What is crazy is that the average person in our society holds this same view, accepting everything placed before them, rather than using sober guidance to analyze the health of the activity and impact to their moral reality.

With every passing day, our popular culture erodes, becoming calloused and hardened against those precious things God deems important. For example, when I first saw a homeless person begging for spare change, my heart went out to him; I wanted to help anyway I could. This shows God's preprogramming of my heart. However, as I grew older, I slowly became desensitized to such feelings. Now, later in life, I see homeless people frequently asking for money. Inappropriately, I have felt myself getting mad at them, rather than helping. As with the seven-year-old girl growing accustomed to violence, by not acting on the desire God places on my heart, I lose the ability to even have such feelings in the future.

A friend of mine grew up in India amidst throngs of panhandling by children and adults. To survive in that culture, he forced away feelings of compassion. Today, he simply feels nothing toward the poor and homeless other than an academic, non-emotional, sorrow. Rather than reaching out in love and compassion, seeing them makes him upset by ruining a moment of seeming peace. How should we respond? While in many cities, begging has become a trade for those who could do more, what do you think would happen if we responded to them in love, or with a pure conscience in a way consistent with our initial internal programming? While giving them our spare change may or may not be appropriate,[229] how would they respond if we invited them into our home for a meal or shower? I have done this repeatedly. While surprisingly many do not accept my invitation, some do. I do this simply to serve. However, I am embarrassed to say that I typically get a peace and satisfaction of unity that outweighs any gift or hospitality. When I do this, I respond in a way that is consistent with a fraction of the grace and love that God shows me.[230] It also helps me better understand God and draws me closer to Him.

K. Power in the Name

Regardless of pronunciation or language translation and specific word used, there is power in the heavenly realms in simply proclaiming the name of Jesus. The *Bible* shares stories of miraculous healing in the name of Jesus Christ. Peter healed a cripple,[231] and the other apostles and followers of Jesus followed suit. Healing the sick and cleansing those afflicted by evil

---

[229] A very close relative was addicted to heroin and ended up overdosing and dying. He was able to continue his habits due to the fact that people gave him money on the street. Had he not been given money, he may have still died, but it would not have likely been from heroin. He would have been forced to make decisions that might have helped him recover, rather than allowing him to continue down a path of escapism.

[230] This idea gives meaning to Jesus' words, "Whoever finds his life will lose it, and whoever loses his life for My sake will find it." Matthew 10:39

[231] "Taking him by the hand, [Jesus] helped him up, and instantly the man's feet and ankles became strong" (Acts 3:7).

spirits were such frequent events, the *Bible* references them generically as, "The apostles performed many miraculous signs and wonders among the people."[232] The power of God manifested itself such that even the disciples' garments healed people.[233]

Even today, whether we need healing, relief from spiritual oppression, or seek to express joy and praise, saying "Jesus" manifests power. Whether pronounced *'JEE-zuhs'* or *'hay-SOOS,'* or in a non-English tongue (from Иисус to يسوع), if the person uttering the term does so with pure intent, our Creator releases some of His authority and power. This is consistent with any term denoting the character of Jesus, whether *'Christ,' 'Jehovah,'* etc. Declaration of His name rebukes evil spirits, as referenced in the *Bible*, and unites us with a spiritual connection to our Creator. It pulls a spiritual reality and force into our physical proximity. If sincere in the intent, it works every time... try it. It is especially effective in confronting deceptions of spiritual oppression and lies told to our spirit. When we feel lost or overwhelmed by feelings or events, proclaim *'Jesus'* aloud to feel the clouds of spiritual oppression part.

L. Pizza

Do we all have the same definitions, whether pizza, any other noun, or adjective? When I became a Christian, at age twelve, it was because I had fun at church and because the more I learned, the more I knew it to be

---

[232] "The apostles performed many miraculous signs and wonders among the people" (Acts 5:12).

[233] "God did extraordinary miracles through Paul, so that even handkerchiefs and aprons that had touched him were taken to the sick, and their illnesses were cured and the evil spirits left them" (Acts 19:11).

true. For the first time in my life, I felt the presence of God in the church we started to attend. However, what brought me back to Christ was the realization that there is a physical tangibility to the forces battling God; innate evil exists. In brief, in Christ, I never really knew real pain or fear. When I hit college, I felt drawn out of curiosity toward illusory satanic readings and the occult. While I believed in God, I did not believe in hell or eternal punishment. I found the whole notion of demons absurd yet interesting. It is here, in this brief period of my life, that my eyes were opened to the reality of spiritual warfare.[234]

True fear suffocates and warps both objective reality and subjective reality into an elusive thirst that can never be quenched. The more I delved, the more I learned that this fear is personified. Hopefully, few reading this book know what I mean. Had it not been for God's amazing grace, I could have drowned in this sea of deception. I once had a book on the role-playing game, *Dungeons & Dragons* that had one of my favorite pictures of a demon named *Asmodeus*.[235] The head of the global church of satan supposedly lived in Boulder, where I was living at the time, so I hoped to learn more about the whole notion. I found a local library and began reading all the books I could find on Satanism and the occult. Amidst this search and reading one afternoon, I suddenly became aware of a presence near me. I looked around, but could not see anything; I felt an overpowering hatred. It was so strong, my legs buckled from underneath

---

[234] "For our struggle is not against flesh and blood, but against the rulers, against the authorities, against the powers of this dark world and against the spiritual forces of evil in the heavenly realms" (Ephesians 6:12).

[235] References to Asmodeus can be found in the book of Tobit (3:8), an Old Testament book only found in the Catholic and Orthodox Bible translations but is not found in the mainstream Bible.

me, and I fell to the floor. It was seething at me and the sensation was palpable. It was bizarre. Perhaps, it was all in my mind, but a series of events followed which I will not go into here, and I left the library. Fortunately, a few weeks later, the Holy Spirit grabbed hold of my heart, I repented, and I was ushered into a new level of relationship with Christ.

I felt God's love for me and clung to it. In focusing on Christ, I was given a genuine joy that flowed out of me like a river. Not understanding what I am saying? When living in northern Spain, some friends knew I loved pizza and took me to a local pizza restaurant. They were so excited to present me with this round, dry, biscuit of an attempt at pizza. Its thin, cardboard-like tasteless crust held a bare sprinkling of hardly discernable toppings, a few pinches of cheese, and dabs of bland sauce. However, out of their experience, this is how my friends knew *pizza*. Unsurprisingly, none of them really liked pizza, as it was based upon this definition, or understanding, of the term. One may never know good pizza, until one tries truly *good* pizza.

When I think of a truly tasty pizza, I think of a chewy sourdough crust with a crispy bottom and edges. The sauce abounds, flashing a rich taste of oregano, basil, and garlic. Thick low-fat mozzarella holds together layers of vegetables that are cooked, but crisp with several types of meat. One slice can be a meal. This is completely unlike the pizza known by my Spanish friends. They had never tried anything close to my definition of *pizza*.

Jesus is the best *'pizza'* we will know. No matter how we imagine the toppings, sauce, cheese, and crust, we cannot tangibly understood it, until our first bite. So it is with a relationship with Christ. We cannot know, until we take our first step. To quote Robert Frost, "And I, I took the road less traveled by, and that has made all the difference."

As explained in Galatians, the byproduct of a life in Christ is "love, joy, peace, patience, kindness, goodness, faithfulness, gentleness, and self-control." [236] Ultimately, what good is anything if it does not bring peace, joy, or love?

*   *   *

In this part, we delved into the identity of Jesus. We learned this definition is mutually exclusive between Christians, Jews, and Muslims, despite the shared history and acceptance of the *Tanakh* between all three. We saw that spiritual law requires a sacrifice, and God is relational and personal, sending Jesus in human form to die for our sins, that we might restore fellowship with our Creator. For those that accept Christ and profess the Truth, Jesus gives grace through His sacrificial death. However, a day is coming when no more grace will be given, just judgment and justice. This is why Scripture continually warns us to 'overcome' [237] and 'run the race to win'.[238]

As we consider the possibility that the *Bible* is true and Jesus really is God in the flesh, it is natural to begin asking the question, "What should I do about it?" In Part V, we explore the next steps for someone who believes Jesus Christ is the God of Creation.

---

[236] "So I say, live by the Spirit, and you will not gratify the desires of the sinful nature" (Galatians 5:16).

[237] In the book of Revelation, chapters 3 & 4 state this seven times.

[238] "I press on toward the goal to win the prize for which God has called me heavenward in Christ Jesus" (Philippians 3:14).

Part V

Living the Life Intended

Living the Life Intended

An investigation of *Intelligent Design* leads to a Designer, an Entity who created not only the Universe, but all life, including humankind. This story delves deeper, with a love so great that this Entity speaks to humanity throughout its history, manifesting Himself in the form of a man. This man lived on the Earth and His words and actions transformed history. This man was Jesus. Jesus, in His non-human form, created everything in the Universe. Although this sounds like an outlandish fairy tale, science and history confirm its truth across a multi-dimensional landscape encompassing physical, moral, and spiritual realities. Since He cares about us enough to know the number of hairs on our heads,[239] it should go without saying that we would want to know Him on a personal level.

A. First Step – Drawing Closer

> *Come to Me, all you who are weary and burdened, and I will give you rest. Take My yoke upon you and learn from Me, for I am gentle and humble in heart, and you will find rest for your souls. For My yoke is easy and My burden is light. (Matthew 11:28-30)*

---

[239] "And even the very hairs of your head are numbered" (Matthew 10:30). See also Luke: 12:7.

It is not enough to simply do the right thing.[240] Many non-Christians lead better 'Christian lives' than some Christians. Unfortunately, unless we accept God and seek Christ, Scripture shares that we will not taste heaven. We must act in Truth,[241] accepting the grace provided by Christ through His death on the cross and knowing the heart of God, in order to join Him in the resurrection. Jesus is the point; we must choose[242] to accept Him, a personified, intelligent *Force*, whose makeup created the brilliance of the Universe, or choose not to accept His grace, and seek a life trying in vanity to find our life. This choice is ours. This is the question; this is the debate; this is our 'free will'.

One day, we will die. It could be today, or it could be tomorrow. There is no guarantee and there should be no expectation that we will live beyond any particular point. It is vanity that we feel entitled to live to a ripe old age. Likely, it will occur sooner than we think or prior to our being 'ready'. When this happens, we will give an account of our life. Are we ready?

Once God lets us to get to a point where we realize that we cannot do life on our own, and we realize our need for the assistance of a higher power, it is quite simple to draw closer. All it takes is a sincere heart crying out to Jesus, proclaiming Him as the focus of our life,[243] earnestly seeking

---

[240] "We who are Jews by birth and not 'Gentile sinners' know that a man is not justified by observing the law, but by faith in Jesus Christ" (Galatians 2:15-16).

[241] "So we, too, have put our faith in Christ Jesus that we may e justified by faith in Christ and not by observing the law, because by observing the law no one will be justified" (Galatians 2:16) and, "Yet a time is coming and has now come when the true worshipers will worship the Father in Spirit and Truth, for they are the kind of worshipers the Father seeks. God is Spirit and his worshipers must worship in Spirit and in Truth" (John 4:23).

[242] "But if serving the Lord seems undesirable to you, then choose for yourselves this day whom you will serve…" (Joshua 24:15).

[243] "If you confess with your mouth, 'Jesus is Lord,' and believe in your heart that God raised Him from the dead, you will be saved" (Romans 10:9).

to turn away from all the ways we placed either ourselves, or worldly things, above Him. We then receive His perfect forgiveness.[244] When this proclamation is sincere and genuine, there may not be lightening or some supernatural proclamation, but rest assured, slam dunk, our salvation is secured, and we embark upon a new journey. We get a clean slate, with sincere forgiveness for all the ways we have put the created above our Creator. Regardless of where we are in life, from the darkest reaches of the occult to the embrace of a church pew, He is there, instantaneously, when we seek Him. No circumstance is too great for God to overcome on our behalf. He loves us with our past, whatever it may be. He loves us not for what we have done, but for who we will be in Him, through His grace.[245] He leverages our past, our pains, and our growth. When we search for God and draw near to Him, He opens the door for us to see all things new, and we realize He is already with us.[246]

As with most relationships, the best way to improve our relationship with God is by spending time with Him. We accomplish this by coming to Him daily through (1) reading and meditating on His Word, as laid forth in the *Bible*, (2) praying (praise, petition, and listening) to Him in all things, (3) fellowshipping or spending time with other like-minded individuals through community and life, and (4) obeying what the Holy Spirit reveals to us as Truth and living in belief. Whatever He reveals to us will be

---

[244] "If we confess our sins, He is faithful and just and will forgive us our sins and purify us from all unrighteousness" (1 John 1:9).

[245] "But because of His great love for us, God, who is rich in mercy, made us alive with Christ even when we were dead in our transgressions- it is by grace you have been saved… it is the gift of God" (Ephesians 2:4-5, 8).

[246] "Submit yourselves, then, to God… Come near to God and He will come near to you" (James 4:7-8).

consistent with His Word, in looking at and understanding Scripture in a contextual whole. In order to understand this perspective, it is important to study *The Bible*, delving into its simple, yet "living" complexities on a continual basis.[247] As a result, journaling these events with the Lord may remind us of His hand in our lives upon reflection. [248]

B. Living Life to Honor Him

So whatever you do, whether in word or deed, do it all in the name of the Lord Jesus, giving thanks to God the Father through Him. (Colossians 3:17)

Once we make this sincere and heartfelt proclamation of God, embodied in Christ, as the focus of our lives, it becomes our responsibility to learn more about how He wants us to live. With salvation secured, the purpose of life during our short moments on Earth is to live a life that honors Him. We accomplish this by simply following what Christ reveals to us when we seek Him.[249] He may change our lives immediately, or He may have us work through certain issues or relationships over time. What is important to remember is that we are now a part of His plan. We no longer

---

[247] "For the word of God is living and active. Sharper than any double edged sword, it penetrates even to dividing soul and spirit, joints and marrow. It judges the thoughts and attitudes of the heart" (Romans 4:12).

[248] Because when we pray, we are to, "go into a room and pray to a God who is unseen," these times of devotion, prayer, and praise are frequently called 'quiet times' even though we may be doing everything from yelling, singing, or being completely still or prostrate on the ground.

[249] "Trust in the Lord with all your heart and lean not on your own understanding; in all your ways acknowledge Him, and He will make your paths straight" (Proverbs 3:5-6).

seek to find our lives, but to lose them in Him.[250]

Our actions show the place of our heart, so if we have things in our life that bind us (ie: drugs, alcohol, greed, sex, or other addictions), rather than trying to fix these symptoms, we should focus on our heart and underlying motivation. To keep this focus, and especially during times of struggle, it becomes critical that we take every thought captive[251] to determine the direction of our focus,[252] whether on Jesus, the Perfecter of our faith,[253] or on the created and the things of this world.

As we love others as we love ourselves and live out a life that denies our own interests over the interests of Christ, we discover that God reveals our life and true *raison d'etre*, or reason for being. Thus, we live life in a state of constant refinement in a fascinating adventure. As we seek Him, our impurities are boiled away. He disciplines us as children and leads us into deeper Truths. There is no promise of money or any worldly gain other than to never have need for worry and that each day will take care of itself. With the fulfillment of God in us, we realize that this life is just a form of illusion or test, so our need for materiality offers no eternal benefit.

I used to consider myself a 'selfish Christian,' meaning that I was only a Christian, because I got something out of it. I received love, peace,

---

[250] Luke 9:23-27

[251] "We demolish arguments and every pretension that sets itself up against the knowledge of God, and we take captive every thought to make it obedient to Christ" (2 Corinthians 10:5).

[252] Submit yourselves, then to God. Resist the devil, and he will flee from you. Come near to God and He will come near to you. Wash your hands, you sinners, and purify your hearts, you double-minded" (James 4:7-9).

[253] "...let us throw off everything that hinders and the sin that so easily entangles. And let us run with perseverance the race marked out for us, fixing our eyes on Jesus, the pioneer and perfecter of faith" (Hebrews 12:1-2).

direction, and a proactive anointing on my life, where I felt His protection and blessing. However, as my time in Christ moves from years to decades, I learn that there is so much more. While I may always get something out of my relationship with Christ,[254] it is not always something that I want at that point in time. He states that He disciplines us and refines us, as a smith refines the impurities out of gold with fire.[255] I, for one, continue to receive lots of refining and discipline. Some days, I really do not want to be melted or chiseled. It hurts! However, time, wisdom, and careful reflection shown these times in my past to be needed refinement to help me grow more like Christ in the future. Praise be to our God that we can trust His ways, knowing without worry that He looks out for us and prepares a way for those who earnestly seek Him.

C. Refocusing our Priorities

I know someone with a nice new car who was always uptight about getting a scratch on it, stressed about where they parked, bitter when anything happened to it, and abrupt and rude to others about taking care of it. This person was me. I replaced it with an older, yet reliable car that still has some personality, and I have yet to worry about it at all. I am happy to loan it to friends or people in need. By fleeing from the materialism associated with my car, I found greater peace and happiness, not to mention more mind share to redirect toward more positive aspects of life.

---

[254] "If you remain in Me and My words remain in you, ask whatever you wish, and it will be given to you" (John 15:7).

[255] "These have come so that your faith – of greater worth than gold, which perishes even though refined by fire – may be proved genuine and may result praise, glory, and honor when Jesus Christ is revealed" (1 Peter 1:7).

In the same way, we may gain more by turning control of our lives over to our Creator and seeking to grow in relationship with Him. In focusing on faith in God's Word and His Truths, I hold firm, not giving sway to emotion or circumstance.

We represent fragile and weak clay pots, containing treasure. Nothing is special about the pot, allowing the focus on the treasure, which is Christ.[256] God is bigger than my weaknesses, and my imperfections do not diminish His deity. If God can use me, He can use you. Our ability to clear away the things that keep us from hearing Him and being still, becomes tantamount that we listen to His soft voice and take steps where He wants us to go.

A couple years ago, my startup company had yet to produce material revenue, and I poured my savings into growing the company. With ongoing spending and no income, I found myself running toward a financial wall. In praying about my circumstances many times, I felt God's call to simply continue my course. However, I saw a deadline approaching where my savings would end, meaning an end to cash flow. This wall approached quickly. Yet, instead of fearing the outcome, I got excited. I felt God's peace that my path was in the right direction, so I just dug into this faith, with great anticipation for what God would do. At the last minute, another opportunity popped up, giving me a term of steady income, allowing me to stabilize my finances. Such 'walls' hit us throughout life, so it is imperative that we focus on our faith and not our emotions or circumstances.[257] When

---

[256] "But we have this treasure in jars of clay to show that this all-surpassing power is from God and not from us" (2 Corinthians 4:7). New King James Version.
[257] "Since we have that same spirit of faith, we also believe and therefore speak..." (2 Corinthians 4:13). "Therefore we do not lose heart. Though outwardly we are wasting away, yet inwardly we are being renewed day by day" (verse 16). "So we fix our eyes not on

we pursue God, He opens a door.[258]

D. Innocence

How do our actions and environment change our trajectory and impact our views and perceptions in life? Not only do we find consequences for the things we say and do in physical reality, but we need to be keenly aware of the moral consequences and spiritual implications. We need to guard our hearts and protect our innocence.[259] In our brief time on this planet, we become desensitized to many things, and, as we do, we lose pieces of our innocence. This *jading* and *loss of purity* then impacts our actions and life. A few years ago, my five-year-old nephew watched *Spiderman* and was terrified by the violence for days, yet now, by age eight, it is one of his favorite films. He does not hesitate at the same, or even greater, levels of violence. Repeated exposure caused him to become immune to the violence. This type of desensitization takes something precious from us, keeping us from empathizing both in watching films and in life.

Similarly, when we continue to subject ourselves to increasing levels of violence, what we once deemed as gruesome horror films will eventually seem acceptable. I have many friends, in fact, that love horror films and actively seek them out. In this case, we desensitize ourselves to violence and horror. This issue manifests in other aspects of our lives, too. Sex is another example, where the more exposure people experience, they must

---

what is seen, but on what is unseen, since what is seen is temporary, but what is unseen is eternal" (verse 18).

[258] "Jesus looked at them and said, 'With man this is impossible, but with God all things are possible'" (Matthew 19:26).

[259] "Above all else, guard your heart, for it is the wellspring of life" (Proverbs 4:23).

push the boundaries farther for the same desired result. Similarly, with the consumption of alcohol or drugs, or by shopping, or in seeking material possessions, as well as a variety of other pursuits, we slowly build a tolerance requiring us to push prior states of seeming, yet illusory, contentment. As we do this, our hearts grow callous and even cold, pushing us further from what we seek. When this happens, we see spiritual reality, where the illusive focus of our pursuit shows the spiritual battle in which we dwell. Such unmet desires should serve as warning and caution us to refocus on Jesus.

Today, society promotes a view that wisdom comes from every type of new experience. Perceptions of 'right' and 'wrong' have become politically incorrect, and every experience being deemed equal to every other. There is little distinguishing the form, or impact, of the experience. I often hear questions like, "How do you know if you like it or not, if you have never tried it?" This approach is not wise, and it can harm us physically, as well as spiritually. Wisdom suggests that rather than experiencing something ourselves, we can look at precedents previously set by other people to then make our own conscious conclusion on whether or not we want to experience similar outcomes. We must take responsibility for our actions and deem each according to the wisdom and expectation that we will live with the outcomes. Rather than taking heroin, or a similar highly addictive drug, to see if we like it, we can first study the chemical effects, social reactions, and addictive impacts. From this analysis, we may make a valid and informed decision to abstain from taking the drug without every trying it. Our culture surprisingly objects to this basic wisdom.

While it may be fun to simply act and be, escaping immediate accountability and responsibility, our actions frequently come back to impact our lives in the most tangible of ways. Sometimes this can be beneficial, but oftentimes, it hurts. I hear people say, "Everything happens for a reason." I tend to agree but often, the result stems from our own stupidity with the result being something God wanted us to avoid. Thus, we may be better off protecting ourselves in some areas, working to abstain from being desensitized, allowing our hearts and minds to be raw to the world around us. A friend of mine started dating a guy and got pregnant. She felt troubled about the child, the timing, and the situation, blaming God for her situation. God gave us rules to follow, not to be cruel or to keep us from having fun, but to provide context for things like having a child. God loves her and the child. While He can turn the situation around to bless all involved, He likely prefers us to avoid such situations in the first place.

At times, we must fight to protect our eyes. In doing so, we also protect our hearts, working diligently to battle the natural tendency of our bodies and minds to succumb to potentially harmful influences in the world around us. In doing so, we remain open to a higher law, an objective reality that seeks to work through us and shape us for our ultimate purpose.

When my niece was little, her working parents placed her into an educational daycare. This took place in the cultural melting pot of San Francisco. I remember picking her up one day and seeing her playing with children of other races with parents from other countries, religions, and cultures. However, her parents did a good job, and society had yet to teach her to even care about such differences. Now, she is older and very socially conscious, trying to erase the same distinctions she did not see years prior and treat people as if she did not see distinctions of race, religion, socioeconomic tier, and culture. In the *Bible*, Jesus tells His disciples that unless they can live and act like little children, they have no place in His Kingdom of Heaven.[260] This example speaks to the innocence and openness of children, accepting others in humility. There is a reality that transcends our subjective and limited understanding of the world in which we live.

The gravity of perfection, the pursuit of purity, and the corruption of sin reveal themselves throughout history. In the exodus of Moses from Egypt, only two of the original Jewish Egyptians lived their faith in God, so He waited for the rest of the Egyptian generation to die off before carrying the new generation into the Promised Land. Similarly, to protect the purity and innocence of His people in the Promised Land, God called for the destruction of many nations that rebuked Him. It is not that God condones killing, but rather that He is willing to go to extreme lengths to protect purity. We need to acknowledge and be aware of this spiritual truth and its associated perspective on reality. Additionally, God warned the Jews to not give into the traditions and practices of the people that remained, but rather to be a beacon of light before all nations and testify to their relationship with Him.

---

[260] "At that time the disciples came to Jesus and asked, "Who, then, is the greatest in the kingdom of heaven?" He called a little child to him, and placed the child among them. And he said: "Truly I tell you, unless you change and become like little children, you will never enter the kingdom of heaven. Therefore, whoever takes the lowly position of this child is the greatest in the kingdom of heaven. And whoever welcomes one such child in my name welcomes me" (Matthew 8:1-5).

This spiritual reality reveals the tremendous consequences of a corruption we may not even see, unless we fight to protect our innocence. It places a value on tiny slices of our lives, esteeming the innocent at the expense of the worldly and corrupt. This should give us all pause, as we see the choices and options in our daily lives. There may be stark spiritual consequences. We must fight to hold our innocence and not throw it away or lose it down the wide and common path of life. God urges us to trust Him and to love and accept our brothers and sisters, just like little children untainted by nature and society. We may see, but not be overcome, touch, but not taste, experience hatred, yet love, understand greed, yet give, identify corruption, yet forgive and bless, behaving with the innocence and purity of little children before the world around us. We should not tap not into this broken world for our strength, but into our Creator eternal.

E. Spiritual Battle

As we read and watch the news, it becomes critical that we understand that while God is ultimately in control, the spiritual realities of darkness rule everything that is not focused on Christ. Not all things happen for good. Pain and strife abound. Death, famine, destruction, and anguish flourish. We should view events with an eye of discernment based upon spiritual understanding. While Christ can use any negative event toward His higher purpose, a world away from God does not follow His will for humanity. We live in a world filled with extreme anguish, lost hope, and forgotten dreams. This highlights another spiritual reality. Forces of darkness exist and have been granted limited dominion on Earth, with direct manifestation into the physical world.[261] Our reality expands beyond the mere physical before us and carries into the moral and spiritual realms around us. "So we fix our eyes not on that which is seen, but on what is unseen. For what is seen is temporal, but what is unseen is eternal."[262] It reminds us that, "... our struggle is not against flesh and blood, but against the rulers, against the authorities, against the powers of this dark world and against the spiritual forces of evil in the heavenly realms."[263]

F. Judgment Verses Grace: Eternity

Judgment comes for each of us, whether we expect it or not. None of us knows the day or hour of our fate; it could be today, or it could be tomorrow. However, many people seem to hold a vain expectation that they, and everyone they know, deserve to live a long, full, bountiful life. Not only is this inaccurate, but it is grossly misleading. Jesus teaches that each day is a blessing and that our time (or *death*) could come at any moment. When this moment arrives, we no longer hold the luxury of making choices that impact our forthcoming eternity. Our window of opportunity closes, and we must deal with the outcome of our words, actions, and decisions. This realization is sobering. If we cannot grasp a finite Universe, how can we grasp an infinite Eternity? As with the analogy

---

[261] The god of this age has blinded the minds of the unbelievers, so that they cannot see the light of the gospel of the glory of Christ, who is the image of God" (2 Corinthians 4:4).
[262] "So we fix our eyes not on what is seen, but on what is unseen. For what is seen is temporary, but what is unseen is eternal" (2 Corinthians 4:18).
[263] "For our struggle is not against flesh and blood, but against the rulers, against the authorities, against the powers of this dark world and against the spiritual forces of evil in the heavenly realms" (Ephesians 6:12).

of the ball, there are only two choices. There is the mass path of destruction that the *Bible* calls hell, or *Gehenna*,[264] and there is the road less traveled for those few who overcome, called *Heaven*. The choice is up to each one of us.

Still not sure? Then, make every effort to seek Truth; study, research, and pray to God who is unseen that He might be revealed to you. Despite it being glossed over in most social circles, politics, school, and even church, it is your most important decision.

G. Prayer & Transformation

While the previous topics may promote one intellectual curiosity or support one's beliefs, the likelihood of them 'making someone a Christian' is low. After all, I identify as a *Christian* not because of apologetics or simple proofs, but because I know *It* to be true through prayer, the presence of the Holy Spirit, and God's transformation of my life. Additionally, I know what God has done in my life and see His hand clearly weaving throughout it. I feel His presence and peace through prayer and in action upon what He has taught me to be true and right. He continues to transform me, refining my many faults and molding me into a tool that can be better used by Him.

Each follower of Christ has a story, a testimony of how our Creator transformed his or her life. One of the best ways to know if God is real and true is to read His Word, the *Bible*, and meditate on it in prayer. Ask God to

---

[264] Gehenna is sometimes the Greek word that is translated into 'hell'; it was the garbage dump outside Jerusalem, where people put the rotting bodies of the dead and the trash and refuse of the city to burn on a continual basis.

170

reveal greater truths and realities to you, and most importantly, be honest with yourself. Seek to sincerely know Him; otherwise, it is just humankind's vain attempt to reach the divine, and there are already many religions out there. God wants more for us than simply filling pews. He calls us into *relationship* with Him, our Creator.

H. Living Abundantly

With pure faith, God expands each aspect of our lives. Through ongoing relationship with God, we learn His perfect way, not straying to the left or the right. He grants us wisdom, providing discernment on the steps we should take, and clarity on the direction we should go. When we uphold to this path, everything we ask happens as we believe,[265] in faith, it will occur. To grow in knowledge of and intimacy with the Creator of the Universe, we must not only hold wisdom, but act upon it. Even though '*gravity*' pulls us down, we tap into the Source, Christ, continually, whether in times of pain and tribulation or joy and delight. When we do this, it is credited to us as righteousness and builds our faith. In this pursuit throughout our few and precious days, we experience the fullness and abundance of life's original intent.

Apathy and indifference saturate our culture. However, in Christ's strength, we shall overcome. Life holds trials and tribulations, offering opportunity for wisdom, humility, and refinement. The purpose of these trials is to refine our faith, increase our wisdom, and provide experience,

---

[265] "Then He touched their eyes and said, 'According to your faith let it be done to you'; and their sight was restored" (Matthew 9: 29-30).

allowing others to look into our lives. We model the highs and lows in life to others, offering encouragement, as well as bringing glory to our Father in Heaven. He calls us to build a personal relationship with Him with a foundation of commitment built on His terms, not ours. Demanding life on our terms, definitions, and schedule is mere vanity. It is a clear indication of our selfish, immature, and imperfect state.

Jesus explains, "Whoever can be trusted with very little can also be trusted with much, and whoever is dishonest with very little will also be dishonest with much. So if you have not been trustworthy in handling worldly wealth, who will trust you with true riches?"[266] Let us, then, show our trustworthiness out of love, not obligation, to receive an eternal inheritance. Let us store up not things of this world that rot, decay, or rust, but eternal rewards in Heaven.

*   *   *

We live amidst physical, moral, and spiritual realities. In focusing on just a limited scope of our senses, our perception limits our understanding of the world. While vast beyond common comprehension, the Universe shows tremendous balance and design. Science helps us discern and qualify a system of Creation that displays the clear splendor of a Designer. The Universe is an *open-system*, with God interacting throughout its beginning, until its end across physical, moral, and spiritual realities. Humanity is not a

---

[266] "Whoever can be trusted with very little can also be trusted with much, and whoever is dishonest with very little will also be dishonest with much. So if you have not been trustworthy in handling worldly wealth, who will trust you with true riches?" (Luke 16:10-11).

mistake, but meticulously formed into the physical consciousness and unique individuality in each person. Our Designer is not an impersonal force and seeks to know and be known. God, who loves us deeply, limited Himself as man, experiencing our joy, our struggles, and our pain. God embodies His own Creation in Jesus.

Repeatedly, humans attempt to exist in the Universe that was created by God with no interaction or dependency upon Him. However in doing so, despair and anguish eventually overwhelm us, unless we develop and continue in a personal relationship with our Creator. History repeats itself in human folly when we seek to fill our spiritual void with anything but relationship with our Maker. Thus, we seek God with our heart, soul, and mind. It is an emotional faith, a faith of peace, and an intellectual faith.

Churches and centers of worship around the globe focus on finding and seeing moral and spiritual realities, yet frequently find it difficult to see God. We see history repeat itself. While resurgence of faith grows in pockets around the world and at different times in history, few people identify with and pursue a personal relationship with God, the Creator of all things. Gravity pulls on all of us, with spiritual forces battling for our souls in dimensions felt, but not seen. People from Lisbon to St. Petersburg see the steeples of churches in every village and town, yet equate faith in God with nursery rhymes and tales told long ago. God holds every atom in our Universe in balance to foster and grow our individual relationships with Him through our daily lives. "A man reaps what he sows." [267] We must manage both our words and our deeds, aware of not

---

[267] "Do not be deceived: God cannot be mocked. A man reaps what he sows" (Galatians 6:7).

only the physical world, but moral realities and spiritual repercussions. Each of us needs wisdom and focus to venture down the road less traveled.

God is around us. He sees you. He cares whether you have dragged your luggage down the road on a hot day, and He feels the tears on your cheeks. There is more. Listen to the inherent moral coding in your heart that tells you the physical world impacts eternity. It is subject to moral reality, enfolded in spiritual truths which comprise the Universe. God loves each of us and wants us to maximize the value of this life through a relationship with Him. It is our responsibility to learn how He communicates with us and to discern His voice amidst the chaos and the constant barrage of the world around us. Whether it be a mobile device, headphones, or a hectic work schedule, we need to make Him a priority, in order to optimize our steps. This means we must become quiet and still, to listen, to hear, and then, subsequently, to follow.

Our world has laws that govern it. As much as we may want to control and define these laws, for most of us, we have very limited, if any, power to do so. While I am responsible for my decisions, words, and actions, as I focus on Christ, I know that God is ultimately in control. This is our invitation. We choose to either trust Him and have purpose, peace, and love *or* place our faith in the created and lose our life in vanity. There are only two ways to live, and we all have to make the choice. Doubt is human; it is inevitable, but when we find ourselves dismissing arguments and ideas prematurely, we must doubt our doubts. Banging one's head against the wall of life day-after-day is no way to live a fruitful life. While we cannot control physical, moral, and spiritual reality, we can revere our Creator and refuse to succumb to the pressures of the world around us. Our Creator shares:

Fear God and keep His commandments, for this is the whole duty of man. For God will bring every deed into judgment, including every hidden thing, whether it is good or evil. (Ecclesiastes 12:13-14)

Come, you blessed of My Father, inherit the kingdom prepared for you from the foundation of the world: for I was hungry, and you gave Me food; I was thirsty, and you gave Me drink; I was a stranger, and you took Me in; I was naked, and you clothed Me; I was sick, and you visited Me; I was in prison, and you came to Me... inasmuch as you did it to one of the least of these my brethren, you did it to Me. (Matthew 25:34-40)

So do not conform any longer to the pattern of this world, but be transformed by the renewing of your mind. Then you will be able to test and approve what God's will is- His good, pleasing, and perfect will. (Romans 12:2-3)

Therefore love is the fulfillment of the law. (Romans 13:10)

But grow in the grace and knowledge of our Lord and Savior Jesus Christ. (2 Peter 3:18)

Whoever is thirsty, let him come; and whoever wishes, let him take the free gift of the water of life. [268] *(Revelation 22:17)*

---

[268] 'water of life' - which is Christ Jesus

It is only through the grace of God and a heart that seeks Him that we discipline our thoughts, actions, and words to experience fullness, peace, and purpose in our lives. Experiencing our Creator is not about religion or man's pious perfection to approach God. It is the opposite. Instead, the *universal Creator* reaches out to us, wanting us to know, honor, and follow Him, so that we can attain and live in relationship with Him. Our time is brief. This is our calling; it is our responsibility; it is our opportunity. If nothing else, try it out. Read the Bible and praise, petition, and listen to our Creator for fifteen minutes, or more, every day for three months. Watch Him transform your life. As we live it out, we follow a dynamic adventure filled with wonder of meaning, engulfed in love, and directed with purpose. It is your adventure. Live it abundantly.

Appendices

A. Transformation of the Disciples

Each of Jesus' disciples and his means of death:

1. Andrew – crucified
2. Bartholomew – crucified
3. James (son of Alphaeus) – crucified
4. Peter – crucified
5. Philip – crucified
6. Simon – crucified
7. James (son of Zebedee) – sword
8. Matthew – sword
9. Thaddeus – arrows
10. James (brother of Jesus) – stoned
11. Thomas – spear
12. John – natural causes

B. Additional Historical Documents

i. **Elephantine Papyri** – An Aramaic collection of contracts and letters that document Jewish life shortly after they were destroyed in 586 B.C.

ii.   **Enuma Elish** – An Akkadian seven-tablet epic recounting the Genesis creation account, dating from the early second millennium B.C.

iii. **Gilgamesh Epic** – An Akkadian recount of Gilgamesh, the ruler of Urak who meets the only survivor of a great flood, dating from the early second millennium B.C.

iv.   **Jehoiachin Dockets** – Akkadian texts from the reign of Nebuchadnezzar II, as chronicled in Kings 25, dating from the early sixth century B.C.

v.   **Lamentations over the Destruction of Ur** – Sumerian poem mourning the destruction of Ur by the Elamites, as chronicled in Lamentations, dating from the early second millennium B.C.

vi.   **Merneptah Stele** – Egyptian story recounting the Egyptian Pharaoh's, Merneptah, conquest of Jewish Israel, dating from the thirteenth century, B.C.

vii. **Mesha Stele** – Moabite recounting of Mesha, the king of Moab's, rebellion against Israel's king Omri (2 Kings 3:4), dating from the ninth century, B.C.

viii. **Nabonidus Chronicle** – Akkadian recount of King Nabonidus's son's, Belshazzar, rule of the kingdom (Daniel 5:29-30), dated from the mid sixth century, B.C.

ix.   **Nebuchadnezzar Chronicle** – Akkadian text dating from the early sixth century, B.C., that describes the reign of Nebuchadnezzar II, including his siege of Jerusalem in 597 B.C. (2 Kings 24:10-17)

x. **Sennacherib's Prism** – Akkadian text describing Sennacherib's siege of Jerusalem in 701, B.C., and his taking of Hezekiah as a prisoner (2 Kings 19:35-37).

xi. **Shalmaneser's Black Obelisk** – Akkadian stone describing a gift given to Assyria's King Shalmaneser III by Israel's King Jehu, dated in the ninth century, B.C.

xii. **Shishak's Geographical List** – Egyptian description detailing out the Jewish cities he captured during his campaign into Judah and Israel (1 Kings 14:25-26), dated from the tenth century, B.C.

xiii. **Siloam Inscription** – Hebrew inscription by a Jewish worker describing the construction of an underground conduit for Jerusalem's water supply during the reign of Hezekiah (2 Kings 20:20; 2 Chronicles 32:30), dated from the late eighth century, B.C.

C. Required Elements for Human Life[269]

In order to sustain *life-as-we-know-it*, we need an extremely fine-tuned balance of lighter and heavier elements. The smallest differentiation in any one of the following sixty elements would negate life. All are required in precisely the levels shown, with health and viability as a species requiring strict accuracy.

---

[269] Emsley, John, The Elements, 3rd ed., Clarendon Press, Oxford, 1998.

Element	Mass of element in a 70-kg person	Volume of purified element	Element would comprise a cube this long on a side:
Oxygen	43 kg	37 L	33.5 cm
Carbon	16 kg	7.08 L	19.2 cm
Hydrogen	7 kg	98.6 L	46.2 cm
Nitrogen	1.8 kg	2.05 L	12.7 cm
Calcium	1.0 kg	64mL	8.64 cm
Phosphorus	780 g	42mL	7.54 cm
Potassium	140 g	16mL	5.46 cm
Sulfur	140 g	67mL	4.07 cm
Sodium	100 g	10mL	4.69 cm
Chlorine	95 g	6mL	3.98 cm
Magnesium	19 g	10.mL	2.22 cm
Iron	4.2 g	0.5mL	8.1 mm
Fluorine	2.6 g	1.7mL	1.20 cm
Zinc	2.3 g	0.3mL	6.9 mm
Silicon	1.0 g	0.4mL	7.5 mm
Rubidium	0.68 g	0.4mL	7.6 mm
Strontium	0.32 g	0.1mL	5.0 mm
Bromine	0.26 g	64.µL	4.0 mm
Lead	0.12 g	10.µL	2.2 mm
Copper	72 mg	8.0µL	2.0 mm
Aluminum	60 mg	22 µL	2.8 mm
Cadmium	50 mg	5.7µL	1.8 mm
Cerium	40 mg	4.8µL	1.7 mm
Barium	22 mg	6.1µL	1.8 mm
Iodine	20 mg	4.0µL	1.6 mm
Tin	20 mg	3.4µL	1.5 mm
Titanium	20 mg	4.4µL	1.6 mm
Boron	18 mg	7.6µL	2.0 mm
Nickel	15 mg	1.6µL	1.2 mm
Selenium	15 mg	3.1µL	1.5 mm
Chromium	14 mg	1.9µL	1.3 mm
Manganese	12 mg	1.6µL	1.2 mm
Arsenic	7 mg	1.2µL	1.1 mm
Lithium	7 mg	13.µL	2.4 mm
Cesium	6 mg	3.2 µL	1.5 mm

Mercury	6 mg	0.4µL	0.8 mm
Germanium	5 mg	0.9µL	1.0 mm
Molybdenum	5 mg	0.4µL	0.8 mm
Cobalt	3 mg	0.3µL	0.7 mm
Antimony	2 mg	0.3µL	0.7 mm
Silver	2 mg	0.1µL	0.6 mm
Niobium	1.5 mg	0.1µL	0.6 mm
Zirconium	1 mg	0.1µL	0.54 mm
Lanthanum	0.8 mg	0.1µL	0.51 mm
Gallium	0.7 mg	0.1µL	0.49 mm
Tellurium	0.7 mg	0.1µL	0.48 mm
Yttrium	0.6 mg	0.1µL	0.51 mm
Bismuth	0.5 mg	51 nL	0.37 mm
Thallium	0.5 mg	42 nL	0.35 mm
Indium	0.4 mg	55 nL	0.38 mm
Gold	0.2 mg	10 nL	0.22 mm
Scandium	0.2 mg	67 nL	0.41 mm
Tantalum	0.2 mg	12 nL	0.23 mm
Vanadium	0.11 mg	18 nL	0.26 mm
Thorium	0.1 mg	8.5 nL	0.20 mm
Uranium	0.1 mg	5.3 nL	0.17 mm
Samarium	50 µg	6.7 nL	0.19 mm
Beryllium	36 µg	20 nL	0.27 mm
Tungsten	20 µg	1.0 nL	0.10 mm

About the Author

Timm Todd competed as a professional athlete at events around the world and spent twenty years working in both large public companies and a variety of technology startups. His keen sense of analysis and views of the world around us offer readers a novel and insightful perspective of life beyond the superficial. He hails from eight states and three countries, always looking for the next big adventure...

Should you have interest in following the author, he is available on social media at:

Facebook:	https://www.facebook.com/seekingtruth.discussion/
Twitter:	Seeking Truth: @seeking14548048
Website:	http://www.seekingtruth.world
Email:	author@seekingtruth.world

For book inquiries or special bulk ordering and/or communication, please send correspondence to:

Timm Todd
PO Box 4064
Hilo, HI 96720